T0142906

Springer Theses

Recognizing Outstanding Ph.D. Research

Aims and Scope

The series "Springer Theses" brings together a selection of the very best Ph.D. theses from around the world and across the physical sciences. Nominated and endorsed by two recognized specialists, each published volume has been selected for its scientific excellence and the high impact of its contents for the pertinent field of research. For greater accessibility to non-specialists, the published versions include an extended introduction, as well as a foreword by the student's supervisor explaining the special relevance of the work for the field. As a whole, the series will provide a valuable resource both for newcomers to the research fields described, and for other scientists seeking detailed background information on special questions. Finally, it provides an accredited documentation of the valuable contributions made by today's younger generation of scientists.

Theses are accepted into the series by invited nomination only and must fulfill all of the following criteria

- They must be written in good English.
- The topic should fall within the confines of Chemistry, Physics, Earth Sciences, Engineering and related interdisciplinary fields such as Materials, Nanoscience, Chemical Engineering, Complex Systems and Biophysics.
- The work reported in the thesis must represent a significant scientific advance.
- If the thesis includes previously published material, permission to reproduce this must be gained from the respective copyright holder.
- They must have been examined and passed during the 12 months prior to nomination.
- Each thesis should include a foreword by the supervisor outlining the significance of its content.
- The theses should have a clearly defined structure including an introduction accessible to scientists not expert in that particular field.

More information about this series at http://www.springer.com/series/8790

Javier Galego Pascual

Polaritonic Chemistry

Manipulating Molecular Structure Through Strong Light–Matter Coupling

Doctoral Thesis accepted by
Universidad Autónoma de Madrid, Madrid,
Spain

 Springer

Author
Dr. Javier Galego Pascual
Física teórica de la materia condensada
Universidad Autónoma de Madrid
Madrid, Spain

Supervisors
Prof. Francisco J. García-Vidal
Física teórica de la materia condensada
Universidad Autónoma de Madrid
Madrid, Spain

Dr. Johannes Feist
Física teórica de la materia condensada
Universidad Autónoma de Madrid
Madrid, Spain

ISSN 2190-5053   ISSN 2190-5061   (electronic)
Springer Theses
ISBN 978-3-030-48700-3   ISBN 978-3-030-48698-3   (eBook)
https://doi.org/10.1007/978-3-030-48698-3

This Springer imprint is published by the registered company Springer Nature Switzerland AG
The registered company address is: Gewerbestrasse 11, 6330 Cham, Switzerland

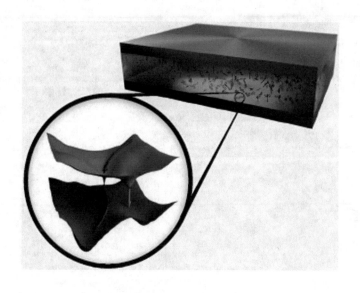

Supervisors' Foreword

One of the most important phenomena in quantum electrodynamics is the so-called "strong coupling" regime, which appears when the interaction between light, i.e., a confined electromagnetic field, and an electronic/vibrational matter excitation, i.e., an exciton, is so strong that the photon and matter components mix to create hybrid light/matter states, called polaritons. This is achieved when the Rabi frequency (energy exchange rate between the exciton and the electromagnetic mode) becomes faster than the decay and/or decoherence rates of either constituent. These polaritons are very attractive from both fundamental and applied perspectives because they inherit properties from their two constituents. Mutual interactions resulting from the matter component lead to non-linear effects, whereas low effective masses, which come from the light constituent, enable new applications such as polariton condensation and low-threshold lasing. In this way, this hybrid character of polaritons has been mainly used to achieve new functionalities in which polaritons are thought of as dressed photons, exploiting exciton–exciton coupling to create interacting photons.

However, after some seminal experiments reported by the group led by Prof. Thomas Ebbesen in the University of Strasbourg in 2012, it has become clear that the strong coupling regime can be used for an alternative purpose to significantly modify internal material properties of organic systems by dressing the excitons with virtual photons. In particular, it has been shown that strong coupling of a macroscopic collection of organic molecules to a Fabry–Perot cavity mode can largely alter the energy landscape of the molecules in such a way that photochemical reactions and even ground-state chemical reactions can be modified. A new area of research has been inaugurated in the last decade: Polaritonic Chemistry. This is precisely the title of the thesis of Dr. Javier Galego, which aims to provide the theoretical and fundamental foundation for this field of investigation.

The main theoretical challenge that we found when we initiated this line of research is that we needed to combine the expertise from, at least, three different subjects: chemistry, quantum optics, and nanophotonics. This is why the work carried out by Javier Galego in this thesis is so valuable: he needed to become familiar with and acquire expertise in three very distinct areas, each of them with its own theoretical and numerical framework. It is remarkable that Javier was able to work successfully in this very interdisciplinary field of research, something that is not very common for a Ph.D. student. This is why we trust that beginners to the field and experts alike will find this thesis useful as Javier Galego has made an extra effort to present the new concepts, techniques and theoretical findings in a very detailed and accessible way.

The first part of the thesis (Chaps. 1 and 2) is intended to introduce both the state of the art in this new field of research and the theoretical background needed to develop the theory of polaritonic chemistry. In line with the interdisciplinary character of this thesis, Chap. 2 is divided into three parts in which the main ingredients taken from nanophotonics ("General light-matter Hamiltonian"), chemistry ("Molecular Hamiltonian"), and quantum optics ("Cavity Quantum Electrodynamics") are introduced in detail.

Once this basic knowledge is summarized, Chaps. 3 and 4 are devoted to analyzing, from a very fundamental and general perspective, how the molecular structure can be altered when the phenomenon of strong coupling emerges. In particular, the powerful concept of "Polaritonic Potential Energy Surfaces" is put forward in Chap. 3, which serves to visualize and understand quantitatively how the energy landscape in which a chemical reaction takes place is modified when polaritons are formed. An important aspect also discussed in these two central chapters is the differences/similarities found between the case of single-molecule strong coupling and the more common situation in which an ensemble of organic molecules is strongly coupled to a cavity electromagnetic mode. In this regard, the relevant phenomenon of "collective protection" is discussed at length in Chap. 4.

The main results and findings of this thesis are discussed in Chaps. 5 and 6. Using simplified models, Chap. 5 demonstrates how a photo-isomerization reaction can be suppressed by taking advantage of strong coupling. On a more positive tone, it also shows how a single photon can trigger a many-molecule reaction thanks to the modification of the energy landscape provided by the phenomenon of collective strong coupling. This example nicely illustrates the idea behind the field of Polaritonic chemistry: as a difference with other approaches to modify chemistry, this procedure is not intended to guide the motion of the nuclear wavepacket along a given path but to completely change and tune the roads in which a nuclear wavepacket can move. Finally, Chap. 6 is devoted to analyzing the most recent experimental findings of modification of ground-state chemical reactions induced by strong coupling. By using the so-called Shin-Metiu model, in which there is one degree of freedom for the nuclei and one for the electron, we introduce the photon as a third ingredient and find the conditions for altering chemical reaction rates in the ground state.

Not all the fundamental questions associated with the brand-new field of Polaritonic Chemistry have been answered in this thesis and there are still many important issues that need further theoretical and experimental insights, in particular the link to the case of cavity-modified ground-state chemistry. It is an exciting time to enter into this new area of research and we hope that this thesis, which summarizes the research carried out by Dr. Galego during his time as a Ph.D. student, will serve both as a helpful introduction and as an inspiring piece of theoretical work.

Madrid, Spain   Prof. Francisco J. García-Vidal
March 2020   Dr. Johannes Feist

Abstract

Polaritonic chemistry is an emergent interdisciplinary field in which the strong interaction of organic molecules with the electromagnetic field is exploited in order to manipulate the chemical structure and reactions of the system. This thesis is devoted to the theoretical study of the internal structure and processes in the organic polaritons that arise in these hybrid light–matter systems. In most theoretical descriptions of the strong coupling regime between light and organic molecules, the latter are treated using simplified descriptions in which the role of the internal nuclear structure is significantly reduced. Our work fully embraces this molecular complexity by combining the usual theoretical descriptions of light found in cavity quantum electrodynamics with the complete molecular characterization used in chemistry, built upon the concept of potential energy surfaces. This leads to the development of a theory in which the tools and concepts of chemistry can be generalized to hybrid light–matter systems.

While in standard chemistry we make a distinction between light and matter, this is no longer true in polaritonic chemistry. The two entities become profoundly mixed, completely altering the properties of the whole. The features of the system are a product of the interaction of the molecule with the electromagnetic vacuum, redefined by the confinement of an optical cavity. Remarkably, the material and chemical properties can be strongly altered even when there is no strong external input of energy. This motivates the field for more experimental and theoretical efforts with the goal of introducing such chemical control in technological applications. In our work, we theoretically study polaritonic chemistry in order to further understand the current experiments and challenge them with new predictions.

In order to build our theory, we first present an overview of the theoretical background necessary of quantum chemistry and cavity quantum electrodynamics. We then combine both descriptions in order to study the molecular structure in strong coupling, analyzing the limits of validity of the Born–Oppenheimer approximation and demonstrating how the cavity induces nuclear correlations between spatially separated molecules. We thus develop a theory of polaritonic chemistry in which we formally study the system for an arbitrary number of molecules in terms of polaritonic potential energy surfaces. Of particular relevance is the study of collective phenomena in strong coupling, which is central in reshaping the energy landscape of hybrid systems. This theory is then applied to two general molecular models that present some form of photochemical process. In the first one, we demonstrate the general suppression of photochemical reactions by influencing the excited-state energy surfaces that govern the dynamics of such processes. Then, in the second molecular model, we prove the possibility of opening novel reaction pathways by smartly manipulating the surfaces based on the theory developed previously. This would enable the possibility of triggering many photochemical reactions over a large number of molecules after absorption of one single external photon, something forbidden in standard photochemistry. Finally, we study the ground-state structural modifications of the light–matter system, investigating the possibility of influencing the reactivity of thermally driven chemical reactions. We demonstrate that quantum electrodynamical effects are indeed able to strongly modify the reactivity in the ground state, observing a collective enhancement for large ensembles of adequately oriented molecules.

List of Publications

1. *Cavity-induced modifications of molecular structure in the strong coupling regime.* **Javier Galego**, Francisco. J. García-Vidal, and Johannes Feist. Physical Review X 5, 041022 (2015).
2. *Suppressing photochemical reactions with quantized light fields.* **Javier Galego**, Francisco. J. García-Vidal, and Johannes Feist. Nature Communications 7, 13841 (2016).
3. *Many-molecule reaction triggered by a single photon in polaritonic chemistry.* **Javier Galego**, Francisco. J. García-Vidal, and Johannes Feist. Physical Review Letters 119, 136001 (2017). (Editor's Suggestion with accompanying Viewpoint in Physics 10, 105 (2017)).
4. *Polaritonic chemistry with organic molecules.* Johannes Feist, **Javier Galego**, and Francisco. J. García-Vidal. ACS Photonics 5, 205 (2018).
5. *Cavity Casimir-Polder forces and their effects in ground state chemical reactivity.* **Javier Galego**, Clàudia Climent, Francisco. J. García-Vidal, and Johannes Feist. Accepted in Physical Review X (2019).
6. *Plasmonic nanocavities enable self-induced electrostatic catalysis.* Clàudia Climent, **Javier Galego**, Francisco. J. García-Vidal, and Johannes Feist. Angewandte Chemie International Edition, 10.1002/anie.201901926 (2019). (Publication not featured in this thesis).

Acknowledgements

I want to thank my two thesis supervisors FJ and Johannes for their support. I consider myself very lucky to have had great scientists as advisors in a project of great interest in the scientific community. Undoubtedly, this has made things easier a lot of time.

Since I attended to his lectures for the first time, over more than six years ago, I always had FJ into great consideration as a scientist and as a person, and I know I have been very lucky to have been his student. Although I never assisted Johannes' lectures, there are plenty of things that I have learned from him, in our numerous meetings (that is, every time I went up to his office in order to bother him with my silly questions). I feel particularly fascinated by his computational abilities, which motivated me since the beginning to improve my programming skills, both in and out of research. Thanks to them I have (hopefully) learned to do real science, not only by always asking myself the correct questions and by presenting my results in an appropriate way, but by taking into account the often forgotten human aspect of science.

I haven't worked just with my two supervisors, but I also have collaborated with my good friend Clàudia, to whom I want to say, that thanks to her never-ending good mood and her passion for science, she made the work of my very last year incredibly more enjoyable.

I am also grateful to the members of the jury of my doctoral thesis, Drs. Hernán Míguez, Rosario González Férez, Peter Rabl, Elena del Valle, and Stefano Corni, as well as Alicia Palacios and Alejandro González Tudela for their availability as reserve members. Thank you for your time and willingness to read this manuscript and attend from different places to be part of the defense.

Madrid, Spain   Javier Galego Pascual
May 2019

Contents

1   **Introduction** 1
   1.1   Motivation 1
   1.2   Strong Light–Matter Coupling 3
       1.2.1   Regimes of Interaction Between Light and Matter 4
       1.2.2   Experimental Strong Coupling Realizations 6
       1.2.3   Strong Coupling with Organic Molecules 11
   1.3   Polaritonic Chemistry: State of the Art 14
       1.3.1   Manipulating Excited-State Processes 15
       1.3.2   Ground State Chemistry in a Cavity 17
   1.4   Summary of Contents 19
   References 20

2   **Theoretical Background** 29
   2.1   General Light–Matter Hamiltonian 29
       2.1.1   Maxwell Equations and Coulomb Gauge 30
       2.1.2   Minimal Coupling Hamiltonian 31
       2.1.3   Dipolar Hamiltonian 34
   2.2   Molecular Hamiltonian 36
       2.2.1   Born–Oppenheimer Approximation 37
       2.2.2   Intermolecular Forces 40
       2.2.3   Chemical Processes 42
       2.2.4   Response to the Electromagnetic Field 45
   2.3   Cavity Quantum Electrodynamics 49
       2.3.1   Electromagnetic Fields in Cavities 49
       2.3.2   Common Theoretical Descriptions 52
       2.3.3   From Weak to Strong Light–Matter Coupling 57
   2.4   Summary of Methods Applied in This Thesis 61
   References 62

3   Molecular Structure in Electronic Strong Coupling   67
   3.1   Introduction   67
   3.2   Single Molecule   68
      3.2.1   Bare Molecule Model   68
      3.2.2   Molecule-Photon Coupling   70
      3.2.3   Absorption   73
      3.2.4   Nonadiabatic Corrections in Strong Coupling   75
   3.3   Two Molecules   78
      3.3.1   Method   78
      3.3.2   Absorption   81
      3.3.3   Nuclear Correlation   82
   3.4   Conclusions   84
   References   85

4   Theory of Polaritonic Chemistry   87
   4.1   Introduction   87
   4.2   Polaritonic Potential Energy Surfaces   87
   4.3   Collective Phenomena: The Supermolecule   90
      4.3.1   Collective Protection   92
      4.3.2   Polaritonic Nonadiabatic Phenomena   94
   4.4   Conclusions   96
   References   97

5   Manipulating Photochemistry   99
   5.1   Introduction   99
   5.2   Suppressing Photochemical Reactions   99
      5.2.1   Single Molecule Dynamics   102
      5.2.2   Collective Suppression   104
      5.2.3   Beyond the Single-Excitation Subspace   107
   5.3   Enhancing Photochemistry   110
      5.3.1   Single Molecule Quantum Yield Increase   112
      5.3.2   Triggering of Many Reactions in Collective Strong
             Coupling   114
   5.4   Conclusions   117
   References   118

6   Cavity Ground-State Chemistry   121
   6.1   Introduction   121
   6.2   Theoretical Model   122
      6.2.1   Cavity Born–Oppenheimer Approximation   123
      6.2.2   Shin–Metiu Model   125
   6.3   Effects of the Cavity on Ground-State Reactivity   127
      6.3.1   Reaction Rates   127
      6.3.2   CBOA-Based Model   129
      6.3.3   Resonance Effects   135

6.4   Modifying Chemistry in Realistic Systems   138
     6.4.1   Multi-mode Cavity: Nanoparticle on Mirror   138
     6.4.2   1,2-Dichloroethane Molecule   140
6.5   Collective Effects   141
6.6   Modifications of the Ground-State Structure   148
6.7   Conclusions   151
References   153

7   General Conclusions and Perspective   157
7.1   General Theory of Polaritonic Chemistry   157
7.2   Applications of Cavity-Modified Chemistry   160
7.3   Ending Remarks   161
References   162

Acronyms

a.u.	Atomic units
BOA	Born–Oppenheimer approximation
CBOA	Cavity Born–Oppenheimer approximation
CQED	Cavity quantum electrodynamics
D	Debye (unit)
DBR	Distributed Bragg reflector
DoF	Degree/s of freedom
EM	Electromagnetic
LSP	Localized surface plasmon
MEP	Minimum energy path
PES	Potential energy surface
PoPES	Polaritonic potential energy surface
QED	Quantum electrodynamics
RWA	Rotating-wave approximation
SC	Strong coupling
SPP	Surface plasmon polariton
TC	Tavis–Cummings
TST	Transition state theory
USC	Ultrastrong coupling

Chapter 1
Introduction

1.1 Motivation

From the first photosynthetic organisms, approximately three billion years ago, to the invention of the laser in the twentieth century, light has had a crucial role in shaping the universe and our lives. Light has always interacted with living organisms, providing energy and information from their environment. Evolution towards complex life forms was rendered possible due to the oxygenation of the Earth produced by early photosynthetic cyanobacterias [1]. In order to survive and thrive, animals need to assimilate information from their surroundings, for which they developed photosensitive cells which later evolved into sophisticated organs: the eyes [2]. Light is the means through which humans see each other and form societies, and people throughout the world and across history have understood its importance.

From the earliest times, philosophers in ancient India and Greece considered the question of light, writing on concepts such as reflection and refraction. Based on some of these texts, in the 11th century the arab scholar Ibn al-Haytham[1] (also known as Alhazen) wrote about optics and formulated precise laws of refraction [3]. During the 17th and 18th centuries, an intense scientific debate arose questioning the nature of light. On one hand, Isaac Newton developed his corpuscular theory, arguing that the straight rays of light demonstrated its particle nature. On the other hand, many of his contemporaries such as Robert Hooke and Christiaan Huygens maintained that light was composed of waves. This was later supported by Thomas Young's double-slit experiment, where wave characteristics such as interference could be seen on light, leading to the general acceptance of its wave nature.

We owe the first great revolution in the study of light to James Clerk Maxwell. By the middle of the 19th century a considerable amount of theoretical knowledge about electricity and magnetism had been gathered. In 1861 Maxwell condensed and

---

[1] As a remark, the controlled experimental testing of his scientific hypotheses is considered the first achievement of the modern scientific method. Because of this, together with his pioneering studies on the behavior of light, he is considered the "father of modern optics".

© The Editor(s) (if applicable) and The Author(s), under exclusive license
to Springer Nature Switzerland AG 2020
J. Galego Pascual, *Polaritonic Chemistry*, Springer Theses,
https://doi.org/10.1007/978-3-030-48698-3_1

corrected it into a set of four equations,[2] and stated that electricity and magnetism are two manifestations from the same substance, and that light is an electromagnetic (EM) wave propagating according to those laws. With this, the corpuscular theory appeared to be completely dead, but soon a new revolution would yet again challenge our perception of reality: quantum mechanics.

In 1900 Max Plank found the solution to the ultraviolet catastrophe related to the radiation of a black body. In his explanation there was one revolutionary assumption: light was emitted and absorbed in discrete packets of energy. In 1905 this same hypothesis was used by Albert Einstein to explain the photoelectric effect. These two events eventually led to the birth of quantum mechanics and its concept of wave–particle duality, as well as to coining the idea of "photon". The quantum theory of light began in the 1920s when Paul Dirac introduced a full quantum description of light and matter [4], laying the foundations of the theory of quantum electrodynamics (QED). This stands as one of the most successful scientific theories in history, and its understanding soon brought a plethora of technological development and applications, such as the laser [5], nowadays a basic tool in medicine, industry, and scientific research among others, or the charge-coupled device (CCD) [6], central for digital imaging.

In the following decades, fundamental research and innovative experimental techniques allowed humanity to efficiently control light and matter at the nanoscale. This lead to the dawn of nanophotonics, which has emerged as a dynamic and prolific research area with the promise of a next generation of photonic devices [7]. Opportunities of avant-garde technology arise thanks to achievements such as superresolution microscopy [8], the discovery of metamaterials [9], improved solar cells [10], and nanolitography [11, 12], to cite just a few. Many of the different areas of nanophotonics have as a common ingredient the manipulation of the electromagnetic field at the nanoscale. Of particular interest to this thesis is the tailoring of EM fields to achieve strong interactions between light and matter. With this it is possible to enter the strong coupling regime, where light and matter become profoundly mixed. The excitations of such a hybrid system do not have a purely material or light nature, but rather they inherit properties of both constituents, giving rise to unusual phenomena. These novel excitations (which often can be understood as emerging quasiparticles) are known as *polaritons*, and constitute a promising pathway towards engineering novel materials [13].

One crucial realization of strong coupling is achieved with organic matter [14]. This has attracted a lot of interest in the last decades due to the possibility of achieving very strong interactions even at room temperature, a limit in which quantum features often are washed away by thermal fluctuations. Furthermore, in these materials strong coupling offers an efficient and elegant pathway to shape the material and chemical properties of organic molecules [15]. The work developed in this thesis constitutes a comprehensive theoretical study of the manipulation of chemical properties and reactions in organic materials. This introductory chapter first sum-

---

[2]Originally Maxwell's equations were composed by 20 different expressions. The simplification to only four equations is credited to Oliver Heaviside.

marizes the fundamentals of light–matter interaction at the nanoscale, reviewing the possible experimental platforms to achieve strong coupling with organic molecules. Then we present a state-of-the-art review of the field which this thesis is focused on: polaritonic chemistry.

1.2 Strong Light–Matter Coupling

The lengthy development of quantum electrodynamics had plenty of difficulties on its path. Possibly one of the most notable ones is the appearance of diverging energies in vacuum, cured by renormalization theory. In simple terms, all measurable parameters of particles that can couple to the electromagnetic field are unavoidably "dressed" by local vacuum fluctuations. This effect produces small corrections in energy levels, first observed by Willis Lamb in the hydrogen spectrum [16]. While these corrections cannot be switched off, they do depend on the electromagnetic environment and can thus be modified by manipulating the distribution of modes upon imposing physical limitations to the field, e.g., by placing mirrors or conductors around the atoms. This was first noted by Purcell [17], who predicted that the rate of spontaneous emission for a nuclear magnetic moment should be enhanced by restricting the number of possible EM modes in a resonant electric circuit to only one strong mode. While the prediction was made for nuclear magnetic moments, the argument is valid for any kind of quantum emitter[3] in resonant cavities. In the consecutive years, several studies followed dealing with spontaneous emission rates in atoms near metallic surfaces. Of particular importance is the study by Casimir and Polder [18], where they discuss how vacuum fluctuations can produce a force between an atom and a conducting plane.

All of this new theoretical interest marked the birth of cavity quantum electro-dynamics (CQED) [19]. In a nutshell, the goal of CQED is to isolate a quantum emitter inside a box so that the effects of the electromagnetic vacuum on the emitter are observable. This can be achieved by increasing the strength of the interaction between light and matter. The light–matter coupling strength is of course a relative concept, and we need to compare it to some energy scale to gauge it. Typically, two different regimes are considered: the weak and the strong coupling regimes. The regime of interaction depends on how large the energy scale of the coupling is com-pared to the decay rate of both the light and matter constituents. In the following, we offer a simple discussion of such interaction regimes, for a more involved analysis see Sect. 2.3.

---

[3] We generalize this to "quantum emitters", which may represent any entity that can absorb or emit light, such as atoms, molecules, quantum dots, nanoparticles, etc.

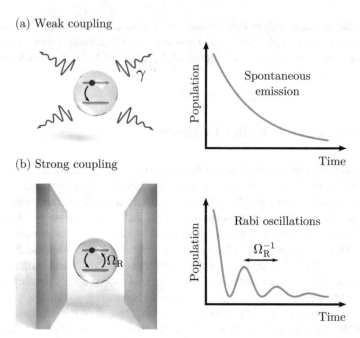

Fig. 1.1 The two main regimes of light and matter interaction in QED **a** weak and **b** strong coupling. Left: conceptual sketches of a single two-level quantum emitter (a qubit, the simplest matter description) in free space, weakly coupled to the EM field, and inside a cavity, strongly coupled to the cavity EM field. Right: time evolution of the population of the excited state of the qubit, showing simple spontaneous emission in weak coupling and Rabi oscillations in strong coupling

1.2.1   Regimes of Interaction Between Light and Matter

When the light–matter energy exchange is slower than the individual decay and dephasing rates (loss of excitation and quantum coherence respectively) of both elements, the system is said to be in the *weak coupling* regime. This is the most common scenario in nature, where the interaction between material (electronic and nuclear) and electromagnetic degrees of freedom can be treated perturbatively [20, 21]. This describes familiar processes such as absorption and emission. The excitation of a quantum emitter has a non-zero probability to be transmitted to the electromagnetic field in the form of a photon (spontaneous emission, see Fig. 1.1a). This is translated in terms of the excited-state lifetime, after which the emitter is said to have emitted a photon. This is typically described by a theory of open quantum systems [22, 23], where the emitter is coupled to a dissipative environment representing the continuum of EM modes that surrounds it. The transition probability depends on the local density of states of the electromagnetic environment of the emitter. Therefore, by placing the emitter inside a resonant cavity or near a conducting surface it is possible to control the emission rate via the so-called Purcell effect mentioned above.

If the relative coupling strength is further increased, the electromagnetic field can no longer be treated perturbatively, and ultimately the system will enter the *strong coupling* regime. Both photons and material excitations have to be treated on equal footing. The system then will be able to coherently exchange energy between both constituents. While typically the exponential decay of the excitations masks this energy exchange as simple emission or absorption phenomena, i.e., the excitation is transferred from the emitter to the electromagnetic field only once, and vice versa, in strong coupling an oscillatory behavior will be observed before loss of excitation, as a rapid series of emission and reabsorption processes (the so-called Rabi oscillations, see Fig. 1.1b). This population exchange between light and matter indicates that photons and material excitations are no longer the proper eigenstates of the system [21]. Instead, new hybrid excitations arise, called polaritons. These states can also absorb and emit light, but at different frequencies than the original emitter, being referred to *upper* and *lower* polaritons, for the larger and smaller energies respectively. The difference in energy is the so-called *Rabi frequency* Ω_R, and corresponds to the oscillation frequency of the emission-absorption cycle between the excited emitter and the photon.

One exceptional feature of strong coupling arises when a collection of emitters interact with the EM field. The entire ensemble collectively interacts with the field and can be understood as a "giant quantum emitter" with a very large dipole moment. The frequency of oscillations is enhanced $\Omega_R = \sqrt{N}\Omega_0$, where N is the number of emitters and Ω_0 is the corresponding single-emitter Rabi frequency. This phenomenon is known as *collective strong coupling*, and is a very common approach to experimentally achieve strong coupling, since coupling strengths of individual emitters are often too weak to be notable. The collective nature of such systems is of utmost importance in many strong coupling effects, as it can correlate emitters that are far away in distance (and therefore not connected) through the EM field. We note that collective strong coupling and polaritons are not an inherently quantum phenomenon, but they arise when electromagnetic modes interact with classical Lorentzian (damped) oscillators, leading also to the $\sqrt{N}$ enhancement when a large number of oscillators are present. Indeed, polaritons appeared first in the context of classical optics as "collective oscillation of polarization charges in the matter" sustained by interfaces that separate media with permittivities of opposite signs [24, 25].

If the strength of the interactions keeps increasing, the system enters the *ultra-strong coupling* (USC) regime, where some additional counter-intuitive effects emerge. For example, the total number of excitations in the system is not conserved, which potentially leads to the global ground state of the system to being dressed by the EM field, even showing purely quantum properties such as squeezing and entanglement [26]. There is no clear agreement on the coupling strength required to consider the system to be in the USC regime, as it heavily depends on the particular system [27–31]. However, signatures typically connected to USC usually appear when the Rabi splitting energy becomes a significant fraction of the transition frequency of the quantum emitter excited state [32].

1.2.2 Experimental Strong Coupling Realizations

Up to now we have discussed the regimes of interaction in a very broad fashion, overlooking the different possibilities to achieve strong coupling in a realistic setup. Polaritons can be achieved in a wide range of systems of various natures, dimensionalities, and energy scales. Experiments can routinely achieve polaritons in solid-state and organic systems, for structures ranging from a few nanometers to milimetric distances, and for microwaves and ultraviolet light. The fundamental purpose or desired technological application is ultimately what determines the experimental realization. For example, some applications may require the device to work in microwave frequencies, such as in the case of superconducting artificial atoms coupled to on-chip cavities [33–35]. Or perhaps we favor the ability of the system to perform at room temperature, for which organic polaritons offer a more suitable platform [36–38]. Both the quantum emitter and the EM mode components of the system present fundamental advantages and restrictions that shape the possibilities for a particular strong coupling realization. In the following, we present these conditions and discuss some examples of possible single-emitter and collective strong coupling systems.

In order to discuss the fundamental limitations of the interaction, it is vital to analyze the nature of the light–matter coupling strength. As we present in detail in Chap. 2, this depends, to a very good approximation, on the electric field amplitude of the system at the position $\mathbf{r}_0$ of the emitter and the dipole moment of the emitter [21]:

$$g(\mathbf{r}_0) = \boldsymbol{\mu} \cdot \mathbf{E}(\mathbf{r}_0). \tag{1.1}$$

There are two main alternatives to effectively increase the coupling strength in order to reach the strong coupling regime.[4] The first is to efficiently choose the right quantum emitters, favoring large dipole moments. Note that in quantum mechanics the dipole moment is an operator, and finding a "large" and "aligned" dipole moment is not necessarily a straightforward task. For example, a quantum emitter may have a very small ground-state permanent dipole, but present a huge transition dipole moment between ground and excited states, making it suitable for strong coupling. The second approach is to engineer cavities that present very large electric field amplitudes. This can be achieved by confining the EM field in very small volumes, as the electric field associated to a EM mode depends on its mode volume as $|\mathbf{E}| \sim 1/\sqrt{V}$. We define this in a proper manner in Sect. 2.3; for now let us focus on the ability of a cavity to concentrate the electric field in very small volumes. Below we review some examples of experimental strong coupling realizations, focusing first on some different cavities presently used to tailor the EM field, and then discussing the variety of possible quantum emitters in which strong coupling is currently viable.

---

[4]While not explicitly listed, increasing the emitter density is often the main approach to achieve strong coupling is some experimental realizations.

(a)   (b)

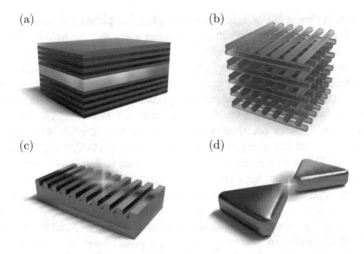

(c)   (d)

Fig. 1.2 Conceptual depictions of some model structures employed to confine light. **a** Fabry–Perot microcavity based on distributed Bragg reflectors. **b** Three-dimensional photonic crystal. **c** Surface plasmon polaritons on a structured metallic surface. **d** Bow-tie nanoantenna hosting a strongly localized surface plasmon resonance

1.2.2.1 Examples of EM Cavities

There are two fundamental approaches to experimentally achieve strong coupling by manipulating the EM environment of the quantum emitters. The first approach is to minimize the losses of the system so that Rabi oscillations can be observed within the lifetime of the cavity and the excited state of the emitter. This is based on a very efficient trapping of light so that a photon inside takes a very long time to exit the system. The second is to localize the light in tiny volumes, thus increasing the electric field amplitude and therefore boosting the light–matter coupling. The two methods are not mutually exclusive; an ideal cavity would incorporate a great confinement of the EM field while trapping light indefinitely (that is, without loss). Some reviews on different kinds of cavities can be found in the literature [39–41]. Let us now overview some examples of cavities that reach the strong coupling regime.

Possibly the simplest structures to achieve strong coupling are the planar microcavities in which two flat mirrors are brought close together so that only a few light wavelengths can fit in between them. The so-called Fabry–Perot microcavity can trap light very efficiently in rather large mode volumes (typically above the diffraction limit, $V \gtrsim \lambda_{EM}^3$, where λ_{EM} is the mode wavelength), which often requires using very large number of emitters to enhance the interaction and achieve measurable Rabi splittings. Depending on the choice of material for the reflectors we can sort between metal and distributed Bragg reflector (DBR) microcavities. The former are easier to fabricate, composed of two parallel layers of a noble metal enclosing the material laterally. However, the fundamental parameters of the metals limit the efficiency of the cavity by introducing losses. This is greatly improved in the case of

DBR microcavities (see Fig. 1.2a), in which the metal planes are replaced by mul-
tilayers of alternating refractive index materials such that for certain wavelength
ranges the reflectivity is close to unity. This offers very large photon lifetimes, even
reaching hundreds of picoseconds [42]. Planar microcavities offer confinement in
only one direction, while in the other two dimensions the EM modes can be arbitrarily extended. Therefore in these cavities photons can be excited with an in-plane
momentum, thus displaying a continuous dispersion relation, which opens a wide
range of possibilities for polariton condensation and superfluidity [43, 44]. A more
intense confinement can be achieved by forming micropillars that exploit total internal reflection. While this greatly increases the losses, it also offers possibilities of
novel devices that can present exotic features such as topological properties [45].

Photonic crystals [46] can be thought of extensions of the DBR structure to two
and three dimensions. By generating a three-dimensional crystal (see for example
Fig. 1.2b) with the appropriate combination of electromagnetic and electronic band
structure, it is possible to rigorously forbid light propagation and scattering inside.
By then creating a defect in this crystalline structure, light states can be confined
without possibility of escaping, leading to the observation of a Rabi splitting [47].
This would theoretically provide one of the most efficient EM field confinement
with tiny losses, however, current experimental realizations have not demonstrated
this yet. Two-dimensional photonic crystals are presently the most promising option
showing great figures of merit [48].

Plasmonic cavities [41] offer a great alternative to achieve strong coupling, offer-
ing sub-wavelength EM field confinement. In here we will consider two types of
cavities that support plasmons of slightly different nature. The first type consists
on engineered material interfaces which support surface plasmon polaritons (SPPs)
[49, 50]. These arise when external light is coupled to the plasmonic excitations of
a metal surface. Due to the momentum mismatch between surface plasmons in the
metallic surface and light in air, these cannot straightforwardly be excited. Instead, it
is possible to shine light passing through a high-refractive-index prism to the metal
surface. Alternatively, it is possible to incorporate an extra wave vector to the system
by devising a surface with a periodic grating [51] (see scheme in Fig. 1.2c). The
quantum emitters located at the surface will be inside the evanescent field of the
plasmonic mode, which can present very high electric field amplitudes. Experiments
of organic materials on top of these systems have led to strong coupling between
SPPs and electronic excitations [52] and nuclear vibrations [53].

Other plasmonic cavities commonly used in strong coupling are based on localized
surface plasmons (LSPs). These cavities exploit the geometric properties of intricate
metallic structures to achieve the best EM field confinement in the literature, however
also showing great losses. Strong coupling has been investigated in a plethora of
different cavities hosting LSPs, such as nanorods [54, 55], nanoprisms [37], and
bow-tie nanoantennas [56–58] (see Fig. 1.2d). Recently the single-molecule strong
coupling limit has been achieved at ambient conditions in the nanoparticle-on-mirror
cavity [38], showing a mode volume for the optically active frequency of $\sim 40\,\mathrm{nm}^3$.
It even has been found that inside the gap of these cavities, atomic-sized defects can
localize LSPs below one cubic nanometer [59].

1.2.2.2 Examples of Quantum Emitters

The choice of quantum emitters heavily relies on the desired properties of the strong coupling realization. As we discussed above, large transition dipole moments favor larger coupling strengths, which lead to an easier observation of the mode splitting. Moreover, the binding energy of the material excitation can also affect the conditions of the experiment, since relatively high temperatures can dissociate excitons with low-energy binding energies (e.g. quantum-well excitons in inorganic semiconductors are typically only supported at cryogenic temperatures). Another important quality of the emitters is the ability to achieve high densities, since strong coupling ultimately depends on $\sqrt{N/V}$. Note that the coupling strength increases as $\sim 1/\sqrt{V}$ and the Rabi splitting with $\sim \sqrt{N}$, therefore it is desirable to fit as many quantum emitters inside the mode volume of the EM field. More parameters that make each quantum emitter unique and potentially more suitable for achieving robust strong coupling are their inertness (i.e., chemical stability) or the possibility of manipulating them in order to fabricate distinct devices. In the following we will review some strong coupling realizations with different types of quantum emitters (Fig. 1.3).

The first experimental observation of Rabi oscillations was made for sodium Rydberg atoms inside Fabry–Perot cavities in the microwave domain [62]. Later, a

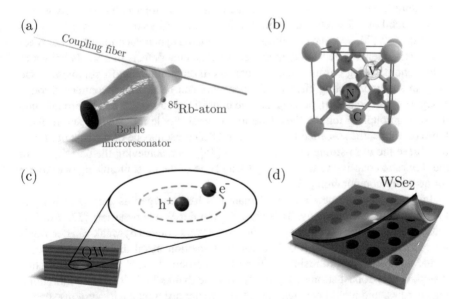

Fig. 1.3 Examples of different quantum emitters. **a** A rubidium atom strongly coupled to a whispering-gallery-mode microresonator, itself coupled to an optical waveguide, as in reference [60]. **b** Simplified atomistic structure of an nitrogen-vacancy center in diamond. **c** DBR microcavity with an inorganic semiconductor quantum well in the center hosting Wannier–Mott excitons (schematically depicted in zoom). **d** Monolayer of WSe_2 (a transition metal dichalcogenide) coupled to a photonic crystal cavity, as in reference [61]

direct observation of the energy splitting in the absorption spectrum was made in an optical cavity [63], achieving for caesium even single-atom strong coupling [64]. This lead to an elegant and sensitive way to detect single atoms and deterministically trap atoms near in the cavity [19, 65]. The great distance between the electron and the nucleus in a Rydberg state makes it possible to have a rather large dipole moment, achieving ~ 1 Debye (D) in these experiments. This makes them highly attractive as single-photon sources [66]. Nevertheless, the convoluted experimental setups and required low temperatures for achieving robust strong coupling heavily restricts the potential of atoms for more sophisticated and practical photonic devices.

The success of CQED in atomic systems quickly brought the attention of the solid-state physics community [67]. The interest was first focused on inorganic semi-conductors, where their intrinsic excitations (excitons) played the role of quantum emitters. These electronic excitations are called Wannier–Mott excitons [68], corre-lated electron–hole pairs, in many ways similar to hydrogen atoms, characterized by very large radius and relatively low binding energies. These states were found to be more stably confined inside quantum wells, quasi-2D regions enclosed by materials of wider bandgap. Solid-state cavity exciton-polaritons were first demonstrated for GaAs quantum wells inside Fabry–Perot microcavities [69], which later led to fasci-nating achievements such as polariton amplification devices [70] and Bose–Einstein condensation [71].

Quantum wells can be further confined into zero-dimensional systems with a set of bound and discrete electronic levels. These "artificial atoms" are known as quantum dots [72, 73], and constitute a central theme in nanotechnology. Other types of artificial atoms have been demonstrated in vacancy defects in crystals known as color centers, being nitrogen-vacancy centers in diamond the most commonly used [74]. In superconducting circuits, Cooper pairs can be quantum confined through Josephson junctions [75], playing the role of artificial atoms that can be brought into the strong coupling regime [34]. Due to the great dipole moment present in these types of qubits, the so-called field of circuit QED presents one of the best platforms to achieve the ultra-strong coupling regime [35], even achieving the best figures of merit in ratio coupling vs frequency [76], and one of the most promising ones to use for quantum computation [33].

In recent years, a new family of materials has emerged as very promising in the fields of nanotechnology. These are the van der Waals materials [77, 78], het-erostructures composed of many atomic monolayers bonded by weak van der Waals interactions. In particular, two-dimensional transition metal dichalcogenides con-stitute a particularly promising platform for photonic devices [79]. In these rising materials very robust strong coupling is possible thanks to their large exciton bind-ing energies, and it has been demonstrated in many different cavity systems such as DBRs [80], photonic crystals [81], and plasmonic structures [82, 83].

Currently one of the most interesting quantum emitters in nanophotonics are organic molecules. Among their numerous advantages they offer high photolumi-nescence quantum yields, very large dipole moments, and great flexibility in the building of photonic devices [14]. Since organic molecules are the main interest of this thesis, in the following we devote an entire subsection to review the field of strong coupling with organic molecules.

1.2.3 Strong Coupling with Organic Molecules

Organic molecules are chemical compounds that contain carbon in their composition. Due to its ability to form chains with other carbon atoms, there is a great variety of different organic molecules, ranging from simple molecules composed of a few atoms (e.g. methane CH_4), to immensely complex molecules such as DNA.

Organic materials still are one of the most interesting platforms to achieve light–matter strong coupling, even more than 20 years after its first realization in an optical microcavity [84]. Such materials present very localized excitations, known as Frenkel excitons [85], characterized by very large binding energies ($\sim$0.1–1 eV) and large transition dipole moments ($\sim$1–5 D), making them optimal for achieving robust strong coupling at room temperature. In general these excitations correspond to excited electronic states bound to single molecules inside the material, thus potentially allowing QED devices with single molecules at room temperature, an ideal scenario for studying quantum optical, nonlinear and saturation effects, such as photon blockade, previously achieved for atoms at cryogenic temperatures [86].

Another interesting quality of organic molecules is their ability to self-aggregate into different types of structures thanks to their weak intermolecular forces. Specifically, molecular aggregates generally present different absorption and emission spectra than the individual molecules they are composed of, potentially red- or blue-shifting the excitation frequency for J- and H-aggregates respectively [87]. Additionally, aggregation can further enhance dipole moments, which made J-aggregates the first class of organic material in which strong emission of polariton states was achieved at room temperature [88]. Due to the wide variety of molecular aggregates it is possible to create narrow absorption spectra tuned to the desired optical or near infrared frequency [89]. This feature is particularly interesting for imitating natural aggregates [90], such as photosynthetic complexes that present very efficient energy absorption and transfer [91]. Notably, strong coupling has been achieved with optically active biomolecules such as β-carotene [92], optical antenna structures in green sulphure bacteria [93], enhanced green fluorescent proteins [94], and reported even in living photosynthetic organisms [95].

Besides aggregates, the attractive van der Waals interactions between molecules can also lead to the formation of well-ordered molecular crystals [96]. In particular, anthracene crystals have been used to achieve strong coupling in the optical regime [97], and even room temperature lasing [36]. The latter achievement was done with a single anthracene crystal, motivated by the belief that strong structural and energetic disorder was the reason previous attempts at lasing did not succeed. However, later experiments demonstrated that it could be accomplished in amorphous small molecule and polymer films [98, 99]. These experiments demonstrated that molecular disorder, intrinsic to many organic material realizations, is not necessarily detrimental for organic polariton device fabrications, as it was previously thought. From a practical standpoint, it is easier to manufacture anything in a disordered state than in an ordered one, making organic materials possibly more advantageous [13].

The versatility of organic molecules has lead to strong coupling experiments in a wide variety of electromagnetic modes such as planar microcavity photons [84, 100–103], surface plasmon polaritons [52, 104–106], surface lattice resonances [107–109], localized surface plasmons [37, 38, 110], and even inside photonic crystals [111]. In the case of localized surface plasmon structures, organic molecules allow to reach strong coupling at the single-emitter level even at room temperature [38, 58, 112], an impressive achievement that promotes the technological development of room-temperature quantum devices. Among other important accomplishments not mentioned above are devices that present polariton–polariton nonlinear interactions [99], nonlinear optical responses [113, 114], and even broadband polariton lasing [115] and polariton-based transistors [116], both at room temperature. Furthermore, strong coupling constitutes a promising solution in material science. It has been demonstrated that it enables the possibility of tuning the work function of organic materials [117], enhancing electrical conductance [118, 119], improving propagation lengths of energy transport (typically of a few nanometers [120]) by several orders of magnitude [121–123], and using organic polaritons to harvest and direct excitations by tuning the cavity mode [124]. Organic systems present also an interesting platform to achieve energy transport between spatially separated molecules [125–128], thanks to nonlocal interactions induced by the cavity.

One unique aspect of organic materials is their internal complexity, apparent in Fig. 1.4a. Typically, organic molecules encompass tens to hundreds of atoms, forming rich structures that involve motion of both electrons and nuclei. Such abundance of degrees of freedom (DoF) opens new pathways for the electronic excitations to

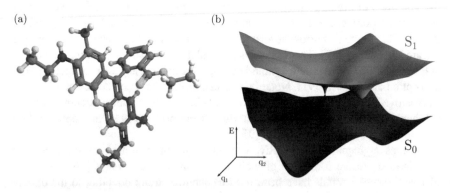

Fig. 1.4 Complexity of organic molecules. **a** Depiction of a rhodamine 6G molecule, commonly used for achieving organic polaritons. **b** Conceptual example of energy dependence with nuclear coordinates q_1 and q_2 of the first two electronic states (S_0 and S_1) of an organic molecule. These, in general, multidimensional hypersurfaces are known as potential energy surfaces. Typical energy landscapes have a dependence on many nuclear degrees of freedom and present multiple electronic states

relax (see an schematic of a typical energy dependence with nuclear coordinates in Fig. 1.4b). For example, the molecule can lose the excitation *nonradiatively*, i.e., without emitting a photon of the exciton frequency, but rather converting the energy into vibrational or rotational motion of the nuclei, i.e., essentially heat. Together with the high level of disorder in organic systems, the rates of dissipation and dephasing become more relevant than in their inorganic counterpart. Furthermore, the interaction between electronic and nuclear DoF (also known as *vibronic coupling*) becomes crucial to explain central features of organic molecules such as the Stokes shift, the difference in energy between absorption and emission spectra.

However, it should be noted that due to the high mass difference between electrons and nuclei, nuclear motion is usually much slower than electronic motion, leading to vibrational modes of lower energy,[5] typically in the mid-infrared spectral region. In some cases, the absorption intensity of certain nuclear bonds is very high, indicating large transition dipole moments [129]. For example, the C=O bond-stretching mode presents a dipole of $\sim$1 D [130] making it suitable for strong coupling. Indeed, this nuclear bond was exploited to achieve strong coupling with infrared modes, first in a polyvinyl acetate polymer [131] and in polymethyl methacrylate [132], even achieving in the latter spatial control over the coupling of vibrations [133]. Additionally, vibrational strong coupling of different molecules and functional groups in the liquid phase was later demonstrated [134].

The acknowledgment of this internal structure led to some pioneering experiments in which the nuclear DoF were exploited. In particular, the structure of some molecules can be altered, which in turn changed the energy of electronic excitation. This allows to turn on and off strong coupling by changing the molecular structure externally and thus detuning the exciton energy from the cavity mode. This was first achieved for a reaction of a porphyrin dye with nitrogen dioxide, which can be reversed through heating of the system [135]. Then, by using the molecule spiropyran, which can undergo reversible change to its isomer merocyanine by externally radiating with UV light [100]. Remarkably, it was shown that in this same setup strong coupling could be used to modify the photoisomerization reaction time from one species to another [136, 137]. Analogous experiments with strong coupling in a perovskite salt demonstrated that the energy barrier of a phase transition could be modified by cavity fields [138]. These experiments demonstrated that the internal structure does not only play a mayor role in organic polaritons, but that it can be exploited to modify the chemistry of a system.

---

[5]Rotational modes have an even lower energy, and are typically not resolved in spectroscopy measurements, being thus reduced to giving fine structure to the vibrational modes. Therefore, usually these two modes are jointly referred to *rovibrational* modes.

1.3 Polaritonic Chemistry: State of the Art

Although the first realization of strong coupling with organic molecules was more than 20 years ago [84], it is only during the last few years that chemical aspects have begun to be explored. Indeed, many experimental works have reported chemical modifications inside cavities [117, 125, 136, 139–142], and much theoretical effort has been devoted to develop an adequate theory of polaritonic chemistry [143–150]. This young—but rapidly growing—field might open the doors to the next generation of polaritonic devices, paving the way towards completely tunable materials whose properties can be controlled for, e.g., optical sensing or energy harvesting applications, among others.

By placing an organic material in a suitable cavity it is possible to bring the system into the strong coupling regime. The molecules and the electromagnetic vacuum are coupled without the need of an external input of energy (as is the case with strong lasers [151]). Thanks to the large dipole moments of organic molecules, it is possible to achieve huge Rabi splittings, completely reshaping the energy structure of the system. This opens the possibility of altering the chemical properties and reactivity of a material, bypassing energy-consuming alternatives such as synthetic material design or control through a large external energy input (e.g., strong lasers or large temperatures).

In strong coupling, the molecules plus the cavity must be thought of a single entity with its own distinct energy levels. It is thus intuitively obvious that this in principle should influence processes that normally take place in the molecular excited state (see in Fig. 1.5 a Jablonski diagram illustrating many of the different process present in organic molecules). Moreover, strong coupling can also have an influence on the electronic ground state of molecules, in two different ways: by

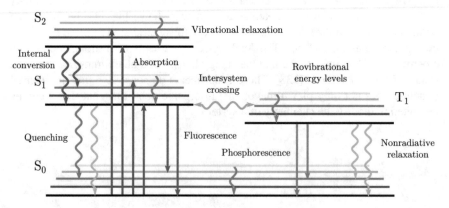

Fig. 1.5 Jablonski diagram depicting the possible processes typically present in organic molecules. Electronic states are denoted by their total spin (S for singlet and T for triplet) and are schematically represented by their rovibrational structure. Straight arrows represent events in which single optical photons are transferred, while wavy arrows depict processes in which energy is transmitted in the form of nuclear motion of the molecule and/or its environment

reaching electronic ultra-strong coupling, where molecule–cavity interactions can potentially "dress" the molecular ground state, and by achieving vibrational strong coupling with ground-state rovibrational states. We therefore separate the discussion of polaritonic chemistry by modifications of electronic excited state chemistry, and modifications of the ground state. In the following we review the experimental and theoretical efforts crucial to the early development of both of these scenarios.

1.3.1 Manipulating Excited-State Processes

The first experimental observation of chemical reactivity being modified in a cavity was done by Hutchison et al. [136] for a photoisomerization reaction. The process was observed in a spiropyran molecule, which undergoes ring opening after UV photoexcitation to form merocyanine, and the inverse reaction is achieved by thermal means. Spiropyran absorbs 330 nm light while merocyanine has an absorption maximum at 560 nm, which is resonant with the Fabry–Perot cavity the molecules are embedded in. Therefore, most of the product molecules are in the strong coupling regime, achieving a Rabi splitting of 700 meV. The authors observe a slow down of the rate of growth of the merocyanine concentration when measured inside the cavity on resonance. The larger the Rabi splitting, the slower the overall reaction is. By altering the energy landscape of the excited-state process, they observed a decrease of the reaction rate. While the system was in the ultra-strong coupling regime (in which the ground state can also be influenced by the cavity), they did not see any change in the thermally-driven back-reaction from merocyanine to spiropyran.

This experiment sparked many theoretical studies aiming to understand this phenomenon. At the time, most existing theoretical models were based on oversimplified descriptions, treating organic molecules as two-level systems. The presence of a more complex internal structure was generally ignored. However, some models took this into account by means of an open quantum systems theory (e.g., Lindblad theory [152]), that is, by assuming that rovibrational modes act like a thermalized bath that induces decay and dephasing on the molecular excitations. The most sophisticated descriptions explicitly treated single vibrational modes as harmonic oscillators around the equilibrium configuration. This model, the so-called Holstein–Tavis–Cummings model [153–155], was early used by Herrera and Spano to predict an enhancement of intramolecular electron transfer in collective strong coupling [156]. The authors discuss the mechanism of polaron decoupling, in which the electronic–nuclear interactions vanish in the thermodynamic limit. This model is a good approximation when the system is close to the equilibrium, which is decidedly not the case in an excited-state chemical reaction where the initial and final nuclear configurations are so different.

Strong coupling is a phenomenon typically studied from the point of view of quantum optics, a field of research that emphasizes the use of simple descriptions to study highly controllable systems. Organic polaritons were often viewed as a means to modify light, and little attention was paid to the intrinsic material properties. The first theory that embraced the complexity of organic systems with the aim to study molecular modifications in strong coupling was developed in [143], one of the studies that we focus on Chap. 3 of this thesis. In this work we aim for a microscopic description of the molecules, fully including their nuclear degrees of freedom. Because of the difficulty of such a task, we treated simple model molecules which could be fully solvable, and analyzed the validity of the Born–Openheimmer approximation, widely-used in chemistry. A related approach was soon after used by Kowalweski et al., in which they analyzed the important nonadiabatic dynamics that emerge in the single-excitation subspace in strong coupling [157, 158]. This method can be interfaced with state-of-the-art quantum chemistry approaches, achieving great accuracy and low computational cost without sacrificing the description of all the internal degrees of freedom [159].

An additional theoretical work was made by Flick et al. [147], where they analyzed matter–photon interactions from the point of view of a quantum-electrodynamical density-functional theory [160]. They demonstrate the potential of this powerful idea to calculate chemical quantities such as bond lengths, nonadiabatic couplings, or absorption spectra. The main challenge of this approach is finding suitable functionals that describe electron–nucleus–photon interactions based on the electron–photon density. It is also of great importance to this thesis (see Chap. 6) the cavity Born–Oppenheimer approximation [148], one of the possible adiabatic approximations that can be performed in an electron–nucleus–photon system. More ideas related to the quantum-electrodynamical density-functional theory were later further explored [161, 162], including additional insight into the intramolecular charge and energy transfer mechanics in strong coupling [163].

Up to now most of the microscopic descriptions mentioned above treated in somewhat detail the electronic and nuclear degrees of freedom such that no more that one or a few molecules could be considered simultaneously due to the exponential complexity of such computational task. Nevertheless, despite the potential of organic molecules, nowadays most strong coupling realizations consist on huge number of emitters. In this context, we extended the theory developed in [143] so that we can treat macroscopic number of molecules in terms of the concept of polaritonic potential energy surfaces [146], a generalization to light–matter system of the ubiquitous potential energy surfaces of chemistry. This theory is part of the focus of Chap. 4 of this thesis. Based on this, we published a theoretical work [164] in which a large collection of photoisomerizable molecules were studied. In particular we introduced a model that represented molecules such as stilbene, azobenzene, or rhodopsin, and studied the single-molecule dynamics and the energy landscape for collective strong coupling. In this study, presented in detail in Chap. 5, we predict a suppression of the reaction that grows more effective with the number of molecules. This effect is a generalization to any kind of energy landscape of the polaron decoupling effect described in [156]. Another collective effect is described in [165] (see Chap. 5), where we discuss the possibilities in polaritonic chemistry of opening new reaction pathways, previously not possible in standard chemistry, without relying on very specific conditions, such as in the case of singlet fission processes [166, 167].

The potential of this theory has been demonstrated by treating big molecular systems using well-known approaches such as QM/MM (quantum mechanics/molecular mechanics) [168, 169]. This accurate technique allows the simulation of realistic experiments while providing detailed insight at the atomistic level. Such method naturally includes nonradiative processes that contributes to the loss of excitation of the molecules, and spontaneous emission of the cavity photon can be straightforwardly added. These processes are often very important in strong coupling with organic molecules, and thus are incorporated in some other descriptions that do not treat explicitly the molecular complexity. For instance, despite it only treating electronic states close to the equilibrium, the aforementioned Holstein–Tavis–Cummings model has been used to theoretically predict polariton-assisted singlet fission [170].

Let us address more recent experiments dealing with polaritonic chemistry of the electronic excited state. One important landmark was achieved recently by Munkhbat et al., in an experiment demonstrating suppression of photobleaching of organic molecules [141]. In this process, a molecule can transfer its excitation from the singlet to the long-lived triplet state (see Fig. 1.5). In this state there is a higher probability of reacting with the atmospheric triplet oxygen ($^3O_2$), leading to chemically unstable species that can damage the photo-active organic molecules [171]. In this experiment it was demonstrated that because of the cavity hybridization of the singlet state, this inherited the short lifetime of the plasmonic modes it was coupled to. This significantly reduced the population transfer to the triplet state, which is the first step of this detrimental process, therefore strongly suppressing the overall photobleaching reaction. Another similar experiment was achieved for the polymer P3HT in a Fabry–Perot cavity, where a threefold reduction of molecular photodegradation is observed [172]. Finally, we note the possibility of using polaritonic chemistry to manipulate the so-called reverse intersystem crossing, that is, the transfer from triplet to singlet states, which has been studied in some experiments and recently discussed [173, 174].

1.3.2 Ground State Chemistry in a Cavity

Most of the research of polaritonic chemistry up to now has been devoted to influencing excited-state reactions and structure via electronic strong coupling. Despite the big relevance of these processes, most common chemical reactions occur in the electronic ground state and are triggered by thermal fluctuations, i.e., the energy contained in the internal motion of the participating molecules is used to overcome the transition state of a reaction. The difference in energy between the reactant state and the transition state is known as activation energy or energy barrier, and its manipulation is one of the main challenges in modern chemistry, for example, by applying external mechanical forces [175] or electric fields [176]. In the context of cavity-modified chemistry, the modification of the ground-state energy barrier was first analyzed for electronic ultra-strong coupling. In the original work of Hutchison et al., the ground-state back-reaction from merocyanine to spiropyran is thermally activated, but the authors did not observe any modification in strong coupling [136]. Indeed, subsequent theoretical studies confirmed that even in the ultra-strong coupling regime for electronic transitions, the ground-state effects are on the order of the single-molecule coupling, i.e., they are not influenced by collective strong coupling [31, 143].

More recently, a number of experiments reported changes in ground-state chemical reactivity, not by exploiting the usual electronic strong coupling, but by tailoring cavities that couple to the desired molecular vibrations [139, 140, 142, 177]. The first observation in 2016 by Thomas et al. reported up to a $\sim$5-fold decrease of the reaction rate of a alkynylsilane deprotection process by strongly coupling the Si–C stretching mode to a infrared cavity [139]. Following experiments achieved strong coupling catalysis, i.e., increments in the reaction rate. First, by achieving ultra-strong coupling with the O–H stretching mode in water, rate increments of 10^2 and 10^4 were achieved for two different hydrolysis reactions [140]. Then, by inducing strong coupling in a C=O bond, present both in the reacting molecule and the surrounding solvent, an increase of the reaction rate of over one order of magnitude has been reported [177].

An experiment of particular relevance is achieved again by Thomas et al. [142]. In this work the authors aim to recover the idea of "mode-selective chemistry" that was so prominent in the 1980s. The original idea was to externally excite specific infrared vibrational modes in order to induce thermally-drive chemical reactions [178]. However, the abundance of rovibrational states at thermal energies that competed with the selected mode made the realization of this idea only feasible at cryogenic temperatures, where relaxation processes were minimized. In this recent experimental study the branching ratio between two different products is modified when the system is in vibrational strong coupling. Not only the reaction rate is modified, but the final outcome of the reaction is changed inside a cavity. It should be emphasized that all of these experiments take place in the dark; there is no explicit input of energy, other than the intrinsic temperature of the sample.

At the time of writing of this thesis, current theoretical approaches do not explain these experiments, and many question remain unanswered. The work in [149] constitutes the first attempt for a microscopic description of ground-state reactivity in strong coupling. This theory, which is the focus of Chap. 6, explores the chemistry of ground-state CQED from a fundamental point of view, studying the formally exact quantum reaction rates and the widely-used transition state theory of chemistry in the context of strong light–matter interactions. Some predictions of this theory are discussed in detail in [179], where quantum chemistry methods are used to simulate realistic reactions in a cavity. In these works we find that the mechanisms that allow to influence the chemistry of the system are related to Casimir–Polder forces and do not explain the resonant condition that the experiments discussed above all share. More recently, a study by Angulo et al. [180] analyzed a particular ground-state charge transfer reaction in vibrational strong coupling. The reactant and product states are modeled as harmonic oscillators so that it is possible to generalize the widely-used Marcus theory to chemical species in vibrational strong coupling. This theory predicts an increase of the charge transfer rate that is most prominent under resonant conditions. However, this is a very specific model in which the reaction rate cannot possibly be slowed down, contrary to the original experiment of 2016. Therefore there is still a need to develop a satisfactory theory of molecule–cavity systems that successfully describes the mechanisms by which chemical reactions can be altered in the ground-state, so we can predict unusual phenomena and further design experimental realizations of interest.

1.4   Summary of Contents

This thesis explores from a theoretical point of view the field of polaritonic chemistry and in general the modification of molecular structure in strong coupling. It is written so that most concepts are supported with the appropriate theoretical background. In the following we explain in more detail the structure of this thesis.

In Chap. 2 we lay the fundamental theoretical background on which the thesis rests upon. We start by providing the crucial ingredients to understand the quantization of the electromagnetic field from Maxwell equations and the Lorentz force, aiming to achieve a quantum electrodynamical Hamiltonian that includes both light and matter. Then, we focus on the material part of this Hamiltonian and overview the theoretical tools used to treat it, such as the Born–Oppenheimer approximation, upon which most modern chemistry is built. Then, we go back to the light–matter Hamiltonian and focus on the possible treatments when the electromagnetic component is confined to a cavity. We overview different theoretical descriptions for cavity QED, such as the ubiquitous Tavis–Cummings model. Finally, we formally introduce the weak and the strong coupling regime based on a simple model, showing the key features of this phenomena.

Next, Chap. 3 is devoted to analyze from first principles the molecular structure in electronic strong coupling. In order to do this we exploit the concepts that we learned from previous cavity quantum electrodynamics models and try to combine them with the molecular description based on the Born–Oppenheimer approximation. We study the effects of strong coupling on the nuclear structure of two different molecules, rhodamine 6G and anthracene, which are reproduced through simplified theoretical descriptions. In particular, we focus on the validity of this approximation, discussing the nonadiabatic terms introduced by the photonic degree of freedom. We compare the absorption spectra for these molecules, with and without approximation, for one photonic mode strongly coupled to one and two molecules. In the case of two molecules, we analyze the nuclear correlations induced by the cavity in both the polaritonic and dark states. The results of this chapter have been published in Physical Review X [143].

The Chap. 4 is devoted to the theory of polaritonic chemistry. We formally introduce the molecular description developed previously into a proper CQED theory. We develop the concept of polaritonic potential energy surfaces, which generalizes the ubiquitous potential energy surfaces of chemistry to hybrid light–matter systems. We discuss this theory, analyzing the physical consequences of such description. In particular we consider the effects of collective strong coupling, which are crucial to understand polaritonic chemistry. These results were published in ACS Photonics [146].

In Chap. 5 we use the theory of polaritonic chemistry to study novel effects of strong coupling in photochemistry. In particular we study the suppression of a model photoisomerization reaction thanks to the hybridization between molecules and photons in a cavity. We present how this effect is remarkably enhanced in the case of collective strong coupling, leading to an almost complete suppression of the reaction. Additionally, we study another model molecule which after photoabsorption can isomerize to a different configuration with a quantum yield of less than unity. We then demonstrate how by tuning the cavity parameters, an increase of the reaction efficiency to essentially one can be achieved. Furthermore, we show how in the case of collective strong coupling this can lead to a succession of isomerization reactions of many molecules, one after another, by originally radiating the system with a single photon. With this we establish the potential of the delocalized nature of polaritons, achieving even the breakdown of the second law of photochemistry. The results of this chapter have been published in Nature Communications [164] and in Physical Review Letters [165].

Finally, in Chap. 6 we introduce the problem of influencing thermally-driven chemical reactions in the ground state. We study the formally exact quantum reaction rates of a model system, in which can apply the cavity Born–Oppenheimer approximation. We develop a theory that allows to explain and predict non-resonant energetic and structural changes to molecules coupled to a quasistatic cavity (e.g., metallic structures that can host plasmonic modes). We then validate our theory by applying it to realistic cavity and molecular systems. We furthermore study the orientation-dependent collective enhancement of the effect both for the reaction rates and the nuclear structural changes. We discuss how our theory can directly connected to well-known van der Waals forces, and more generally, to Casimir–Polder interactions.

References

1. Holland HD (2006) The oxygenation of the atmosphere and oceans. Philos Trans R Soc B Biol Sci 361:903
2. Dawkins R (1997) Climbing mount improbable. WW Norton & Company
3. Nature milestones: photons. Nat Mater 9:S1 (2010)
4. Dirac PAM (1981) The principles of quantum mechanics, vol 27. Oxford University Press
5. Maiman TH (1960) Stimulated optical radiation in Ruby. Nature 187:493
6. Boyle WS, Smith GE (1970) Charge coupled semiconductor devices. Bell Syst Tech J 49:587
7. Koenderink AF, Alu A, Polman A (2015) Nanophotonics: shrinking light-based technology. Science 348:516
8. Pohl DW, Denk W, Lanz M (1984) Optical stethoscopy: image recording with resolution $\lambda/20$. Appl Phys Lett 44:651
9. Smith DR, Padilla WJ, Vier D, Nemat-Nasser SC, Schultz S (2000) Composite medium with simultaneously negative permeability and permittivity. Phys Rev Lett 84:4184

10. Kojima A, Teshima K, Shirai Y, Miyasaka T (2009) Organometal halide perovskites as visible-light sensitizers for photovoltaic cells. J Am Chem Soc 131:6050
11. Schift H (2008) Nanoimprint lithography: an old story in modern times? a review. J Vac Sci Technol B Microelectron Nanometer Struct Process Meas Phenom 26:458
12. Li L, Gattass RR, Gershgoren E, Hwang H, Fourkas JT (2009) Achieving $\lambda/20$ resolution by one-color initiation and deactivation of polymerization. Science 324:910
13. Sanvitto D, Kéna-Cohen S (2016) The road towards polaritonic devices. Nat Mater 15:1061
14. Michetti P, Mazza L, La Rocca GC (2015) Strongly coupled organic microcavities. In: Zhao YS (ed) Organic nanophotonics, nano-optics and nanophotonics, vol 39. Springer, Berlin, Heidelberg
15. Ebbesen TW (2016) Hybrid light-matter states in a molecular and material science perspective. Acc Chem Res 49:2403
16. Lamb WE, Retherford RC (1947) Fine structure of the hydrogen atom by a microwave method. Phys Rev 72:241
17. Purcell EM (1946) Spontaneous emission probabilities at radio frquencies. Phys Rev 69:674
18. Casimir HBG, Polder D (1948) The influence of retardation on the London-van Der Waals forces. Phys Rev 73:360
19. Haroche S, Raimond JM (2010) Exploring the quantum: atoms, cavities, and photons
20. Louisell WH, Louisell WH (1973) Quantum statistical properties of radiation, vol 7. Wiley, New York
21. Cohen-Tannoudji C, Dupont-Roc J, Grynberg G (1997) Photons and atoms
22. Davies EB (1976) Quantum theory of open systems
23. Weisskopf V, Wigner E (1930) Berechnung der natürlichen Linienbreite auf Grund der Dirac-schen Lichttheorie. Zeitschrift für Physik 63:54
24. Mills DL, Burstein E (1974) Polaritons: the electromagnetic modes of media
25. Mills DL, Agranovich VM (1982) Surface polaritons: electromagnetic waves at surfaces and interfaces. North-Holland Publishing
26. Auer A, Burkard G (2012) Entangled photons from the polariton vacuum in a switchable optical cavity. Phys Rev B Condens Matter Mater Phys 85
27. Ciuti C, Bastard G, Carusotto I (2005) Quantum vacuum properties of the intersubband cavity polariton field. Phys Rev B Condens Matter Mater Phys 72:1
28. Ciuti C, Carusotto I (2006) Input-output theory of cavities in the ultrastrong coupling regime: the case of time-independent cavity parameters. Phys Rev A At Mol Opt Phys 74
29. Scalari G, Maissen C, Turčinková D, Hagenmüller D, De Liberato S, Ciuti C, Reichl C, Schuh D, Wegscheider W, Beck M, Faist J (2012) Ultrastrong coupling of the cyclotron transition of a 2D electron gas to a THz metamaterial. Science 335:1323
30. George J, Chervy T, Shalabney A, Devaux E, Hiura H, Genet C, Ebbesen TW (2016) Multiple Rabi splittings under ultrastrong vibrational coupling. Phys Rev Lett 117:153601
31. Martínez-Martínez LA, Ribeiro RF, Campos-González-Angulo J, Yuen-Zhou J (2018) Can ultrastrong coupling change ground-state chemical reactions? ACS Photonics 5:167
32. Kockum AF, Miranowicz A, De Liberato S, Savasta S, Nori F (2019) Ultrastrong coupling between light and matter. Nat Rev Phys 1:19
33. Blais A, Huang R-S, Wallraff A, Girvin SM, Schoelkopf RJ (2004) Cavity quantum electro-dynamics for superconducting electrical circuits: an architecture for quantum computation. Phys Rev A 69:062320
34. Wallraff A, Schuster DI, Blais A, Frunzio L, Huang R-S, Majer J, Kumar S, Girvin SM, Schoelkopf RJ (2004) Strong coupling of a single photon to a superconducting qubit using circuit quantum electrodynamics. Nature 431:162
35. Niemczyk T, Deppe F, Huebl H, Menzel EP, Hocke F, Schwarz MJ, Garcia-Ripoll JJ, Zueco D, Hümmer T, Solano E, Marx A, Gross R (2010) Circuit quantum electrodynamics in the ultrastrong-coupling regime. Nat Phys 6:772
36. Kéna-Cohen S, Forrest SR (2010) Room-temperature polariton lasing in an organic single-crystal microcavity. Nat Photonics 4:371

37. Zengin G, Wersäll M, Nilsson S, Antosiewicz TJ, Käll M, Shegai T (2015) Realizing strong light-matter interactions between single-nanoparticle plasmons and molecular excitons at ambient conditions. Phys Rev Lett 114:157401

38. Chikkaraddy R, de Nijs B, Benz F, Barrow SJ, Scherman OA, Rosta E, Demetriadou A, Fox P, Hess O, Baumberg JJ (2016) Single-molecule strong coupling at room temperature in plasmonic nanocavities. Nature 535:127

39. Vahala KJ (2003) Optical microcavities. Nature 424:839

40. Kavokin A, Baumberg J, Malpuech G, Laussy F (2008) Microcavities

41. Tame MS, McEnery K, Özdemir Ş, Lee J, Maier S, Kim M (2013) Quantum plasmonics. Nat Phys 9:329

42. Steger M, Liu G, Nelsen B, Gautham C, Snoke DW, Balili R, Pfeiffer L, West K (2013) Long-range ballistic motion and coherent flow of long-lifetime polaritons. Phys Rev B 88:235314

43. Carusotto I, Ciuti C (2013) Quantum fluids of light. Rev Mod Phys 85:299

44. Amo A, Lefrère J, Pigeon S, Adrados C, Ciuti C, Carusotto I, Houdré R, Giacobino E, Bramati A (2009) Superfluidity of polaritons in semiconductor microcavities. Nat Phys 5:805

45. St-Jean P, Goblot V, Galopin E, Lemaître A, Ozawa T, Le Gratiet L, Sagnes I, Bloch J, Amo A (2017) Lasing in topological edge states of a one-dimensional lattice. Nat Photonics 11:651

46. Yablonovitch E (1987) Inhibited spontaneous emission in solid-state physics and electronics. Phys Rev Lett 58:2059

47. Yoshle T, Scherer A, Hendrickson J, Khitrova G, Gibbs HM, Rupper G, Ell C, Shchekin OB, Deppe DG (2004) Vacuum Rabi splitting with a single quantum dot in a photonic crystal nanocavity. Nature 432:200

48. Akahane Y, Asano T, Song B-S, Noda S (2003) High-Q photonic nanocavity in a two-dimensional photonic crystal. Nature 425:944

49. Barnes WL, Dereux A, Ebbesen TW (2003) Surface plasmon subwavelength optics. Nature 424:824

50. De Abajo FG (2007) Colloquium: light scattering by particle and hole arrays. Rev Mod Phys 79:1267

51. Pendry J, Martin-Moreno L, Garcia-Vidal F (2004) Mimicking surface plasmons with structured surfaces. Science 305:847

52. Bellessa J, Bonnand C, Plenet JC, Mugnier J (2004) Strong coupling between surface plasmons and excitons in an organic semiconductor. Phys Rev Lett 93:036404

53. Memmi H, Benson O, Sadofev S, Kalusniak S (2017) Strong coupling between surface plasmon polaritons and molecular vibrations. Phys Rev Lett 118:1

54. Zengin G, Johansson G, Johansson P, Antosiewicz TJ, Käll M, Shegai T (2013) Approaching the strong coupling limit in single plasmonic nanorods interacting with J-aggregates. Scientific Reports 3

55. Wen J, Wang H, Wang W, Deng Z, Zhuang C, Zhang Y, Liu F, She J, Chen J, Chen H et al (2017) Room-temperature strong light-matter interaction with active control in single plasmonic nanorod coupled with two-dimensional atomic crystals. Nano Lett 17:4689

56. Lee B, Park J, Han GH, Ee H-S, Naylor CH, Liu W, Johnson AC, Agarwal R (2015) Fano resonance and spectrally modified photoluminescence enhancement in monolayer MoS2 integrated with plasmonic nanoantenna array. Nano Lett 15:3646

57. Kim MK, Sim H, Yoon SJ, Gong SH, Ahn CW, Cho YH, Lee YH (2015) Squeezing photons into a point-like space. Nano Lett 15:4102

58. Santhosh K, Bitton O, Chuntonov L, Haran G (2016) Vacuum Rabi splitting in a plasmonic cavity at the single quantum emitter limit. Nat Commun 7, ncomms11823

59. Benz F, Schmidt MK, Dreismann A, Chikkaraddy R, Zhang Y, Demetriadou A, Carnegie C, Ohadi H, De Nijs B, Esteban R, Aizpurua J, Baumberg JJ (2016) Single-molecule optomechanics in "picocavities". Science 354:726

60. Junge C, O'shea D, Volz J, Rauschenbeutel A (2013) Strong coupling between single atoms and nontransversal photons. Phys Rev Lett 110:213604

61. Wu S, Buckley S, Schaibley JR, Feng L, Yan J, Mandrus DG, Hatami F, Yao W, Vučković J, Majumdar A et al (2015) Monolayer semiconductor nanocavity lasers with ultralow thresholds. Nature 520:69

62. Kaluzny Y, Goy P, Gross M, Raimond JM, Haroche S (1983) Observation of self-induced Rabi oscillations in two-level atoms excited inside a resonant cavity: the ringing regime of superradiance. Phys Rev Lett 51:1175

63. Raizen M, Thompson R, Brecha R, Kimble H, Carmichael H (1989) Normal-mode splitting and linewidth averaging for two-state atoms in an optical cavity. Phys Rev Lett 63:240

64. Thompson RJ, Rempe G, Kimble HJ (1992) Observation of normal-mode splitting for an atom in an optical cavity. Phys Rev Lett 68:1132

65. Hood C, Chapman M, Lynn T, Kimble H (1998) Real-time cavity QED with single atoms. Phys Rev Lett 80:4157

66. Hennrich M, Legero T, Kuhn A, Rempe G (2000) Vacuum-stimulated Raman scattering based on adiabatic passage in a high-finesse optical cavity. Phys Rev Lett 85:4872

67. Kwek LC et al (2013) Strong light-matter coupling: from atoms to solid-state systems. World Scientific

68. Wannier GH (1937) The structure of electronic excitation levels in insulating crystals. Phys Rev 52:191

69. Weisbuch C, Nishioka M, Ishikawa A, Arakawa Y (1992) Observation of the coupled exciton-photon mode splitting in a semiconductor quantum microcavity. Phys Rev Lett 69:3314

70. Savvidis P, Baumberg J, Stevenson R, Skolnick M, Whittaker D, Roberts J (2000) Angle-resonant stimulated polariton amplifier. Phys Rev Lett 84:1547

71. Kasprzak J, Richard M, Kundermann S, Baas A, Jeambrun P, Keeling JM, Marchetti FM, Szymánska MH, André R, Staehli JL, Savona V, Littlewood PB, Deveaud B, Dang LS (2006) Bose-Einstein condensation of exciton polaritons. Nature 443:409

72. Prasad PN (2004) Nanophotonics. Wiley

73. Reithmaier JP, Sęk G, Löffler A, Hofmann C, Kuhn S, Reitzenstein S, Keldysh L, Kulakovskii V, Reinecke T, Forchel A (2004) Strong coupling in a single quantum dot-semiconductor microcavity system. Nature 432:197

74. Gruber A, Dräbenstedt A, Tietz C, Fleury L, Wrachtrup J, Von Borczyskowski C (1997) Scanning confocal optical microscopy and magnetic resonance on single defect centers. Science 276:2012

75. Bouchiat V, Vion D, Joyez P, Esteve D, Devoret M (1998) Quantum coherence with a single Cooper pair. Phys Scr 1998:165

76. Yoshihara F, Fuse T, Ashhab S, Kakuyanagi K, Saito S, Semba K (2017) Superconducting qubit-oscillator circuit beyond the ultrastrong-coupling regime. Nat Phys 13:44

77. Geim AK, Grigorieva IV (2013) Van der Waals heterostructures. Nature 499:419

78. Basov D, Fogler M, De Abajo FG (2016) Polaritons in van der Waals materials. Science 354:aag1992

79. Mak KF, Shan J (2016) Photonics and optoelectronics of 2D semiconductor transition metal dichalcogenides. Nat Photonics 10:216

80. Liu X, Galfsky T, Sun Z, Xia F, Lin E-C, Lee Y-H, Kéna-Cohen S, Menon VM (2015) Strong light-matter coupling in two-dimensional atomic crystals. Nat Photonics 9:30

81. Zhang L, Gogna R, Burg W, Tutuc E, Deng H (2018) Photonic-crystal exciton-polaritons in monolayer semiconductors. Nat Commun 9:713

82. Liu W, Lee B, Naylor CH, Ee H-S, Park J, Johnson AC, Agarwal R (2016) Strong exciton-plasmon coupling in MoS2 coupled with plasmonic lattice. Nano Lett 16:1262

83. Stührenberg M, Munkhbat B, Baranov DG, Cuadra J, Yankovich AB, Antosiewicz TJ, Olsson E, Shegai T (2018) Strong light-matter coupling between plasmons in individual gold bi-pyramids and excitons in mono-and multilayer WSe2. Nano Lett 18:5938

84. Lidzey DG, Bradley DDC, Skolnick MS, Virgili T, Walker S, Whittaker DM (1998) Strong exciton-photon coupling in an organic semiconductor microcavity. Nature 395:53

85. Frenkel J (1931) On the transformation of light into heat in solids. i. Phys Rev 37:17

86. Birnbaum KM, Boca A, Miller R, Boozer AD, Northup TE, Kimble HJ (2005) Photon block-ade in an optical cavity with one trapped atom. Nature 436:87

87. Saikin SK, Eisfeld A, Valleau S, Aspuru-Guzik A (2013) Photonics meets excitonics: natural and artificial molecular aggregates. Nanophotonics 2:21

88. Lidzey D, Bradley D, Virgili T, Armitage A, Skolnick M, Walker S (1999) Room temperature polariton emission from strongly coupled organic semiconductor microcavities. Phys Rev Lett 82:3316
89. Walker BJ, Dorn A, Bulovic V, Bawendi MG (2011) Color-selective photocurrent enhancement in coupled J-aggregate/nanowires formed in solution. Nano Lett 11:2655
90. Scholes GD (2002) Designing light-harvesting antenna systems based on superradiant molecular aggregates. Chem Phys 275:373
91. Tronrud DE, Wen J, Gay L, Blankenship RE (2009) The structural basis for the difference in absorbance spectra for the FMO antenna protein from various green sulfur bacteria. Photosynth Res 100:79
92. Baieva S, Ihalainen J, Toppari J (2013) Strong coupling between surface plasmon polaritons and β-carotene in nanolayered system. J Chem Phys 138:044707
93. Coles DM, Yang Y, Wang Y, Grant RT, Taylor RA, Saikin SK, Aspuru-Guzik A, Lidzey DG, Tang JKH, Smith JM (2014) Strong coupling between chlorosomes of photosynthetic bacteria and a confined optical cavity mode. Nat Commun 5:5561
94. Dietrich CP, Steude A, Tropf L, Schubert M, Kronenberg NM, Ostermann K, Höfling S, Gather MC (2016) An exciton-polariton laser based on biologically produced fluorescent protein. Sci Adv
95. Coles DM, Flatten LC, Sydney T, Hounslow E, Saikin SK, Aspuru-Guzik A, Vedral V, Tang JK-H, Taylor RA, Smith JM et al (2017) Polaritons in living systems: modifying energy landscapes in photosynthetic organisms using a photonic structure. arXiv preprint arXiv:1702.01705
96. Iodoform C (2000) Ternary compounds, organic semiconductors
97. Kéna-Cohen S, Davanço M, Forrest SR (2008) Strong exciton-photon coupling in an organic single crystal microcavity. Phys Rev Lett 101:116401
98. Plumhof JD, Stöferle T, Mai L, Scherf U, Mahrt RF (2014) Room-temperature Bose-Einstein condensation of cavity exciton-polaritons in a polymer. Nat Mater 13:247
99. Daskalakis KS, Maier SA, Murray R, Kéna-Cohen S (2014) Nonlinear interactions in an organic polariton condensate. Nat Mater 13:271
100. Schwartz T, Hutchison JA, Genet C, Ebbesen TW (2011) Reversible switching of ultrastrong light-molecule coupling. Phys Rev Lett 106:196405
101. Kéna-Cohen S, Maier SA, Bradley DD (2013) Ultrastrongly coupled exciton-polaritons in metal-clad organic semiconductor microcavities. Adv Opt Mater 1:827
102. Gambino S, Mazzeo M, Genco A, Di Stefano O, Savasta S, Patanè S, Ballarini D, Mangione F, Lerario G, Sanvitto D et al (2014) Exploring light-matter interaction phenomena under ultrastrong coupling regime. ACS Photonics 1:1042
103. Liu B, Rai P, Grezmak J, Twieg RJ, Singer KD (2015) Coupling of exciton-polaritons in low-Q coupled microcavities beyond the rotating wave approximation. Phys Rev B 92:155301
104. Dintinger J, Klein S, Bustos F, Barnes WL, Ebbesen TW (2005) Strong coupling between surface plasmon-polaritons and organic molecules in subwavelength hole arrays. Phys Rev B Condens Matter Mater Phys 71:35424
105. Hakala TK, Toppari JJ, Kuzyk A, Pettersson M, Tikkanen H, Kunttu H, Törmä P (2009) Vacuum Rabi splitting and strong-coupling dynamics for surface-plasmon polaritons and rhodamine 6G molecules. Phys Rev Lett 103:1
106. Vasa P, Pomraenke R, Cirmi G, De Re E, Wang W, Schwieger S, Leipold D, Runge E, Cerullo G, Lienau C (2010) Ultrafast manipulation of strong coupling in metal-molecular aggregate hybrid nanostructures. ACS Nano 4:7559
107. Baudrion A-L, Perron A, Veltri A, Bouhelier A, Adam P-M, Bachelot R (2012) Reversible strong coupling in silver nanoparticle arrays using photochromic molecules. Nano Lett 13:282
108. Rodriguez SRK, Feist J, Verschuuren MA, Garcia Vidal FJ, Gómez Rivas J (2013) Thermalization and cooling of plasmon-exciton polaritons: towards quantum condensation. Phys Rev Lett 111
109. Väkeväinen AI, Moerland RJ, Rekola HT, Eskelinen AP, Martikainen JP, Kim DH, Törmä P (2014) Plasmonic surface lattice resonances at the strong coupling regime. Nano Lett 14:1721

110. Schlather AE, Large N, Urban AS, Nordlander P, Halas NJ (2013) Near-field mediated plexcitonic coupling and giant rabi splitting in individual metallic dimers. Nano Lett 13:3281

111. Zhen B, Chua S-L, Lee J, Rodriguez AW, Liang X, Johnson SG, Joannopoulos JD, Soljačić M, Shapira O (2013) Enabling enhanced emission and low-threshold lasing of organic molecules using special fano resonances of macroscopic photonic crystals. Proc Natl Acad Sci 110:13711

112. Liu R, Zhou Z-K, Yu Y-C, Zhang T, Wang H, Liu G, Wei Y, Chen H, Wang X-H (2017) Strong light-matter interactions in single open plasmonic nanocavities at the quantum optics limit. Phys Rev Lett 118:237401

113. Barachati F, De Liberato S, Kéna-Cohen S (2015) Generation of Rabi-frequency radiation using exciton-polaritons. Phys Rev A 92:033828

114. Barachati F, Simon J, Getmanenko YA, Barlow S, Marder SR, Kéna-Cohen S (2017) Tunable third-harmonic generation from polaritons in the ultrastrong coupling regime. ACS Photonics 5:119

115. Sannikov D, Yagafarov T, Georgiou K, Zasedatelev A, Baranikov A, Gai L, Shen Z, Lidzey D, Lagoudakis P (2019) Room temperature broadband polariton lasing from a dye-filled microcavity. Adv Opt Mater 1900163

116. Zasedatelev AV, Baranikov AV, Urbonas D, Scafirimuto F, Scherf U, Stöferle T, Mahrt RF, Lagoudakis PG (2019) A room-temperature organic polariton transistor. Nat Photonics 1

117. Hutchison JA, Liscio A, Schwartz T, Canaguier-Durand A, Genet C, Palermo V, Samor P, Ebbesen TW, Samorì P, Ebbesen TW (2013) Tuning the work-function via strong coupling. Adv Mater 25:2481

118. Orgiu E, George J, Hutchison JA, Devaux E, Dayen JF, Doudin B, Stellacci F, Genet C, Schachenmayer J, Genes C, Pupillo G, Samorì P, Ebbesen TW (2015) Conductivity in organic semiconductors hybridized with the vacuum field. Nat Mater 14:1123

119. Salleo A (2015) Organic electronics: something out of nothing. Nat Mater 14:1077

120. Akselrod GM, Deotare PB, Thompson NJ, Lee J, Tisdale WA, Baldo MA, Menon VM, Bulović V (2014) Visualization of exciton transport in ordered and disordered molecular solids. Nat Commun 5:3646

121. Feist J, Garcia-Vidal FJ (2015) Extraordinary exciton conductance induced by strong coupling. Phys Rev Lett 114:1

122. Schachenmayer J, Genes C, Tignone E, Pupillo G (2014) Cavity enhanced transport of excitons. Phys Rev Lett 114:1

123. Gonzalez-Ballestero C, Feist J, Gonzalo Badía E, Moreno E, Garcia-Vidal FJ (2016) Uncoupled dark states can inherit polaritonic properties. Phys Rev Lett 117:156402

124. Gonzalez-Ballestero C, Feist J, Moreno E, Garcia-Vidal FJ (2015) Harvesting excitons through plasmonic strong coupling. Phys Rev B Condens Matter Mater Phys

125. Zhong X, Chervy T, Zhang L, Thomas A, George J, Genet C, Hutchison JA, Ebbesen TW (2017) Energy transfer between spatially separated entangled molecules. Angew Chem Int Ed 56:9034

126. Garcia-Vidal FJ, Feist J (2017) Long-distance operator for energy transfer. Science 357:1357

127. Sáez-Blázquez R, Feist J, Fernández-Domínguez A, García-Vidal F (2018) Organic polaritons enable local vibrations to drive long-range energy transfer. Phys Rev B 97:241407

128. Du M, Martínez-Martínez LA, Ribeiro RF, Hu Z, Menon VM, Yuen-Zhou J (2018) Theory for polariton-assisted remote energy transfer. Chem Sci 9:6659

129. Stuart B (2005) Infrared spectroscopy. Wiley Online Library

130. Koenig JL (1999) Spectroscopy of polymers. Elsevier

131. Shalabney A, George J, Hutchison J, Pupillo G, Genet C, Ebbesen TW (2015) Coherent coupling of molecular resonators with a microcavity mode. Nat Commun 6:5981

132. Long JP, Simpkins BS (2015) Coherent coupling between a molecular vibration and fabry-perot optical cavity to give hybridized states in the strong coupling limit. ACS Photonics 2:130

133. Ahn W, Vurgaftman I, Dunkelberger AD, Owrutsky JC, Simpkins BS (2017) Vibrational strong coupling controlled by spatial distribution of molecules within the optical cavity. ACS Photonics acsphotonics.7b00583

134. George J, Shalabney A, Hutchison JA, Genet C, Ebbesen TW (2015) Liquid-phase vibrational strong coupling. J Phys Chem Lett 6:1027
135. Berrier A, Cools R, Arnold C, Offermans P, Crego-Calama M, Brongersma SH, Gómez-Rivas J (2011) Active control of the strong coupling regime between porphyrin excitons and surface plasmon polaritons. ACS Nano 5:6226
136. Hutchison JA, Schwartz T, Genet C, Devaux E, Ebbesen TW (2012) Modifying chemical landscapes by coupling to vacuum fields. Angew Chem Int Ed 51:1592
137. Fontcuberta-i Morral A, Stellacci F (2012) Light–matter interactions: ultrastrong routes to new chemistry. Nat Mater 11:272
138. Wang S, Mika A, Hutchison JA, Genet C, Jouaiti A, Hosseini MW, Ebbesen TW (2014) Phase transition of a perovskite strongly coupled to the vacuum field. Nanoscale 6:7243
139. Thomas A, George J, Shalabney A, Dryzhakov M, Varma SJ, Moran J, Chervy T, Zhong X, Devaux E, Genet C, Hutchison JA, Ebbesen TW (2016) Ground-state chemical reactivity under vibrational coupling to the vacuum electromagnetic field. Angew Chem Int Ed 55:11462
140. Hiura H, Shalabney A, George J (2018) Cavity catalysis -accelerating reactions under vibrational strong coupling-. ChemRxiv 7234721
141. Munkhbat B, Wersäll M, Baranov DG, Antosiewicz TJ, Shegai T. Suppression of photooxidation of organic chromophores by strong coupling to plasmonic nanoantennas
142. Thomas A, Lethuillier-Karl L, Nagarajan K, Vergauwe RMA, George J, Chervy T, Shalabney A, Devaux E, Genet C, Moran J, Ebbesen TW (2019) Tilting a ground-state reactivity landscape by vibrational strong coupling. Science 363:615
143. Galego J, Garcia-Vidal FJ, Feist J (2015) Cavity-induced modifications of molecular structure in the strong-coupling regime. Phys Rev X 5:41022
144. Bennett K, Kowalewski M, Mukamel S (2016) Novel photochemistry of molecular polaritons in optical cavities. Faraday Discuss 194:259
145. Herrera F, Spano FC (2018) Theory of nanoscale organic cavities: the essential role of vibration-photon dressed states. ACS Photonics 5:65
146. Feist J, Galego J, Garcia-Vidal FJ (2018) Polaritonic chemistry with organic molecules. ACS Photonics 5:205
147. Flick J, Ruggenthaler M, Appel H, Rubio A (2017) Atoms and molecules in cavities, from weak to strong coupling in quantum-electrodynamics (QED) chemistry. Proc Natl Acad Sci 114:3026
148. Flick J, Appel H, Ruggenthaler M, Rubio A (2017) Cavity Born-Oppenheimer approximation for correlated electron-nuclear-photon systems. J Chem Theory Comput 13:1616
149. Galego J, Climent C, Garcia-Vidal FJ, Feist J (2019) Cavity Casimir–Polder forces and their effects in ground state chemical reactivity. arXiv preprint arXiv:1807.10846
150. Ribeiro RF, Martínez-Martínez LA, Du M, Campos-Gonzalez-Angulo J, Yuen-Zhou J (2018) Polariton chemistry: controlling molecular dynamics with optical cavities. Chem Sci 9:6325
151. Brabec T, Kapteyn H (2008) Strong field laser physics, vol 1. Springer
152. Lindblad G (1976) On the generators of quantum dynamical semigroups. Commun Math Phys 48:119
153. Ćwik JA, Reja S, Littlewood PB, Keeling J (2014) Polariton condensation with saturable molecules dressed by vibrational modes. Epl 105:47009
154. Ćwik JA, Kirton P, De Liberato S, Keeling J (2016) Excitonic spectral features in strongly coupled organic polaritons. Phys Rev A 93:33840
155. Spano FC (2015) Optical microcavities enhance the exciton coherence length and eliminate vibronic coupling in J-aggregates. J Chem Phys 142
156. Herrera F, Spano FC (2016) Cavity-controlled chemistry in molecular ensembles. Phys Rev Lett 116:238301
157. Kowalewski M, Bennett K, Mukamel S (2016) Cavity femtochemistry: manipulating nonadiabatic dynamics at avoided crossings. J Phys Chem Lett 7:2050
158. Kowalewski M, Bennett K, Mukamel S (2016) Non-adiabatic dynamics of molecules in optical cavities. J Chem Phys 144:1

159. Fregoni J, Granucci G, Coccia E, Persico M, Corni S (2018) Manipulating azobenzene pho-toisomerization through strong light-molecule coupling. Nat Commun 9:4688
160. Flick J, Ruggenthaler M, Appel H, Rubio A (2015) Kohn-Sham approach to quantum electro-dynamical density-functional theory: exact time-dependent effective potentials in real space. Proc Natl Acad Sci 112:15285
161. Flick J, Narang P (2018) Cavity-correlated electron-nuclear dynamics from first principles. Phys Rev Lett 121:113002
162. Rivera N, Flick J, Narang P (2018) Variational theory of non-relativistic quantum electrody-namics. arXiv preprint arXiv:1810.09595
163. Schäfer C, Ruggenthaler M, Appel H, Rubio A (2019) Modification of excitation and charge transfer in cavity quantum-electrodynamical chemistry. Proc Natl Acad Sci 201814178
164. Galego J, Garcia-Vidal FJ, Feist J (2016) Suppressing photochemical reactions with quantized light fields. Nat Commun 7:13841
165. Galego J, Garcia-Vidal FJ, Feist J (2017) Many-molecule reaction triggered by a single photon in polaritonic chemistry. Phys Rev Lett 119:136001
166. Walker BJ, Musser AJ, Beljonne D, Friend RH (2013) Singlet exciton fission in solution. Nat Chem 5:1019
167. Zirzlmeier J, Lehnherr D, Coto PB, Chernick ET, Casillas R, Basel BS, Thoss M, Tykwinski RR, Guldi DM (2015) Singlet fission in pentacene dimers. Proc Natl Acad Sci 112:5325
168. Luk HL, Feist J, Toppari JJ, Groenhof G (2017) Multiscale molecular dynamics simulations of polaritonic chemistry. J Chem Theory Comput 13:4324
169. Groenhof G, Toppari JJ (2018) Coherent light harvesting through strong coupling to confined light. J Phys Chem Lett 9:4848
170. Martínez-Martínez LA, Du M, Ribeiro RF, Kéna-Cohen S, Yuen-Zhou J (2018) Polariton-assisted singlet fission in acene aggregates. J Phys Chem Lett 9:1951
171. Levitus M, Ranjit S (2011) Cyanine dyes in biophysical research: the photophysics of poly-methine fluorescent dyes in biomolecular environments. Q Rev Biophys 44:123
172. Peters VN, Faruk MO, Asane J, Alexander R, D'angelo AP, Prayakarao S, Rout S, Noginov M (2019) Effect of strong coupling on photodegradation of the semiconducting polymer P3HT. Optica 6:318
173. Stranius K, Hertzog M, Börjesson K (2018) Selective manipulation of electronically excited states through strong light-matter interactions. Nat Commun 9:2273
174. Eizner E, Martínez-Martínez LA, Yuen-Zhou J, Kéna-Cohen S (2019) Inverting singlet and triplet excited states using strong light-matter coupling. arXiv preprint arXiv:1903.09251
175. Hickenboth CR, Moore JS, White SR, Sottos NR, Baudry J, Wilson SR (2007) Biasing reaction pathways with mechanical force. Nature 446:423
176. Aragonès AC, Haworth NL, Darwish N, Ciampi S, Bloomfield NJ, Wallace GG, Diez-Perez I, Coote ML (2016) Electrostatic catalysis of a Diels-Alder reaction. Nature 531:88
177. Lather J, Bhatt P, Thomas A, Ebbesen TW, George J (2018) Cavity catalysis by co-operative vibrational strong coupling of reactant and solvent molecules. ChemRxiv 7531544
178. Pimentel GC, Coonrod JA et al (1987) Opportunities in chemistry: today and tomorrow. National Academies Press
179. Climent C, Galego J, Garcia-Vidal FJ, Feist J (2019) Plasmonic nanocavities enable self-induced electrostatic catalysis. Angew Chem Int Ed
180. Angulo JCG, Ribeiro RF, Yuen-Zhou J (2019) Resonant enhancement of thermally-activated chemical reactions via vibrational polaritons. arXiv preprint arXiv:1902.10264

Chapter 2
Theoretical Background

This chapter presents the essential theoretical background necessary to explain some of the most important concepts discussed throughout this thesis. The aim is to provide the reader with the basic tools to understand the many fundamental equations and approximations used in the contexts of cavity quantum electrodynamics (CQED) and quantum chemistry. We start by addressing the question of what is the quantum Hamiltonian for the light–matter interaction and illustrating what approximations play an important role in its definition. We then focus on the matter part of the light–matter Hamiltonian in order to provide the best possible description of a complex molecule. In this section we address the Born–Oppenheimer approximation, widely used in molecular and solid-state physics and in quantum chemistry. Additionally, we present the description of different characteristic phenomena of organic molecules such as chemical structure and reactions, and their response to the electromagnetic field. Then, we focus on this last part, discussing the important features of CQED and the different theoretical descriptions that study them. Finally we present the fundamentals of the two different regimes of light–matter interaction: weak and strong coupling.

2.1 General Light–Matter Hamiltonian

This section is devoted to introduce the quantum description of light–matter interaction by determining the appropriate Hamiltonian operator. In here we focus only on the essential ingredients to achieve this; a more detailed description can be found in the literature [1, 2].

© The Editor(s) (if applicable) and The Author(s), under exclusive license
to Springer Nature Switzerland AG 2020
J. Galego Pascual, *Polaritonic Chemistry*, Springer Theses,
https://doi.org/10.1007/978-3-030-48698-3_2

2.1.1 Maxwell Equations and Coulomb Gauge

The first step towards quantization of a system of charged particles with the electromagnetic field is to define the Lagrangian that properly describes the classical equations of electromagnetism, namely, the Maxwell equations and the Lorentz force law:

$$\nabla \cdot \mathbf{E} = \frac{\rho}{\epsilon_0}, \tag{2.1a}$$

$$\nabla \times \mathbf{E} = -\frac{\partial \mathbf{B}}{\partial t}, \tag{2.1b}$$

$$\nabla \cdot \mathbf{B} = 0, \tag{2.1c}$$

$$\nabla \times \mathbf{B} = \mu_0 \mathbf{J} + \mu_0 \epsilon_0 \frac{\partial \mathbf{E}}{\partial t}, \tag{2.1d}$$

$$m_i \frac{\partial^2 \mathbf{r_i}}{\partial t^2} = q_i \left[\mathbf{E}(\mathbf{r_i}) + \frac{\partial \mathbf{r_i}}{\partial t} \times \mathbf{B}(\mathbf{r_i}) \right]. \tag{2.2}$$

For a collection of charged particles, $\rho = \sum_i q_i \delta(\mathbf{r} - \mathbf{r}_i)$ is the charge density and $\mathbf{J} = \sum_i q_i \partial_t \mathbf{r}_i \delta(\mathbf{r} - \mathbf{r}_i)$ is the current density. The constants ϵ_0 and μ_0 are the vacuum electric permittivity and the magnetic permeability respectively. $\mathbf{E}$ and $\mathbf{B}$ are the electric and magnetic fields, in which we omit the spatial and temporal dependence ($\mathbf{E} \equiv \mathbf{E}(\mathbf{r}, t)$) for notational convenience. It is useful to express the electric and magnetic fields in terms of some new variables, the vector potential $\mathbf{A}$ and the scalar potential ϕ:

$$\mathbf{E} = -\frac{\partial \mathbf{A}}{\partial t} - \nabla \phi, \tag{2.3a}$$

$$\mathbf{B} = \nabla \times \mathbf{A}. \tag{2.3b}$$

With these definitions, equations Eqs. (2.1b) and (2.1c) are automatically satisfied, while the remaining two Maxwell equations can now be written as:

$$\nabla(\nabla \cdot \mathbf{A}) - \nabla^2 \mathbf{A} + \frac{1}{c^2} \frac{\partial^2 \mathbf{A}}{\partial t^2} + \frac{1}{c^2} \nabla \frac{\partial \phi}{\partial t} = \mu_0 \mathbf{J}, \tag{2.4a}$$

$$\nabla^2 \phi + \nabla \cdot \frac{\partial \mathbf{A}}{\partial t} = -\frac{\rho}{\epsilon_0}, \tag{2.4b}$$

where $c = (\epsilon_0 \mu_0)^{-1/2}$ is the vacuum speed of light. Therefore, the two equations in Eq. (2.4) with the definitions of Eq. (2.3), together with the Lorentz force law in Eq. (2.2), fully describe classical electromagnetic interactions.

The definition of the vector and scalar potentials given by Eq. (2.3) is not unique, i.e. these equations remain invariant under the *gauge transformations* $\mathbf{A} \to \mathbf{A} + \nabla \chi$ and $\phi \to \phi - \partial_t \chi$, where $\chi(\mathbf{r}, t)$ is any arbitrary function of space and time. This property of the equations of electromagnetism gives us a freedom of choice of the potentials $\mathbf{A}$ and ϕ without altering the underlying physics. Astute gauge fixing can greatly simplify the equations of a particular problem. For our purpose we specify the following condition for the vector potential:

$$\nabla \cdot \mathbf{A} = 0. \tag{2.5}$$

This condition is known as the *Coulomb gauge* and allows to simplify the equations to the following set:

$$\nabla^2 \mathbf{A} - \frac{1}{c^2}\frac{\partial^2 \mathbf{A}}{\partial t^2} = -\mu_0 \mathbf{J} - \frac{1}{c^2}\nabla\frac{\partial \phi}{\partial t}, \tag{2.6a}$$

$$\nabla^2 \phi = -\frac{\rho}{\epsilon_0}. \tag{2.6b}$$

Under this choice, the scalar potential ϕ satisfies the Poisson's equation of electrostatics Eq. (2.6b), and thus it corresponds to the instantaneous electrostatic Coulomb potential $\phi(\mathbf{r_i}) = \sum_{j \neq i} q_j/(4\pi\epsilon_0|\mathbf{r}_j - \mathbf{r}_i|)$.

Furthermore, according to Helmholtz's theorem, we can separate the electric field ($\mathbf{E} = \mathbf{E}_\perp + \mathbf{E}_\parallel$) into transverse ($\mathbf{E}_\perp$) and longitudinal ($\mathbf{E}_\parallel$) components, with zero divergence and zero curl respectively. The Coulomb gauge gives direct physical meaning to each component, since by definition, the vector potential is purely transverse, i.e., $\mathbf{A} = \mathbf{A}_\perp$, and thus the components of the electric field are given by $\mathbf{E}_\perp = -\partial_t\mathbf{A}$ and $\mathbf{E}_\parallel = -\nabla\phi$. We can thus separate Maxwell's equations into transverse and longitudinal sets, where the first describe radiation and retarded interactions via electromagnetic waves, and the second describe instantaneous Coulomb interactions between charges. We can use these considerations to simplify further Eq. (2.6a) and get one single equation for the transversal fields:

$$\nabla^2 \mathbf{A} - \frac{1}{c^2}\frac{\partial^2 \mathbf{A}}{\partial t^2} = -\mu_0 \mathbf{J}_\perp, \tag{2.7}$$

where Helmholtz's theorem has also been applied to the current density $\mathbf{J} = \mathbf{J}_\perp + \mathbf{J}_\parallel$.

2.1.2 Minimal Coupling Hamiltonian

Under the Coulomb gauge we have a clear interpretation of the different sets of Maxwell equations and we have defined the scalar potential ϕ. The expressions in Eqs. (2.2) and (2.7) are enough to describe the light–matter system. Note that these equations depend only on two independent variables, namely $\mathbf{A}$ and $\mathbf{r}_i$. We can thus introduce the following form of Lagrangian:

$$\mathcal{L} = \sum_i \left[\frac{1}{2}m_i\dot{\mathbf{r}}_i^2 + q_i\dot{\mathbf{r}}_i \cdot \mathbf{A}(\mathbf{r}_i) - q_i\phi(\mathbf{r}_i)\right] + \frac{\epsilon_0}{2}\int_V dV \left[\mathbf{E}_\perp^2 + c^2\mathbf{B}^2\right]. \tag{2.8}$$

Note that the fields $\mathbf{E}$ and $\mathbf{B}$, and the potential ϕ, are all functionals of $\mathbf{r}$ and $\mathbf{A}$ and thus this Lagrangian depends on two independent generalized coordinates and their time derivatives, $\{\mathbf{r}_i, \dot{\mathbf{r}}_i\}$ and $\{\mathbf{A}_i, \dot{\mathbf{A}}_i\}$. It can be shown that this Lagrangian recovers the original equations of the light–matter system, Eqs. (2.1) and (2.2), by using the Euler–Lagrange equation.

Now that we defined the Lagrangian and the relevant variables $\mathbf{r}_i$, $\mathbf{A}$ for our system, we construct can the classical light–matter Hamiltonian, defined as $\mathcal{H} = \sum_i \dot{Q}_i P_i - \mathcal{L}$, where the Q_i are the generalized canonical coordinates (i.e., our relevant variables) and $P_i = \partial\mathcal{L}/\partial\dot{Q}_i$ are their corresponding canonical momenta. In this case:

$$\mathbf{p}_i = \frac{\partial\mathcal{L}}{\partial\dot{\mathbf{r}}_i} = m_i\dot{\mathbf{r}}_i + q_i\mathbf{A}(\mathbf{r}_i), \qquad (2.9a)$$

$$\mathbf{\Pi} = \frac{\partial\mathcal{L}}{\partial\dot{\mathbf{A}}} = \epsilon_0\dot{\mathbf{A}}, \qquad (2.9b)$$

where we can actually identify $\mathbf{\Pi}$ as minus the transversal displacement field $\mathbf{D}_\perp = -\mathbf{\Pi}$, due to our choice of gauge where $\mathbf{E}_\perp = -\dot{\mathbf{A}}$. The Hamiltonian of the system thus reads

$$\mathcal{H} = \sum_i \frac{\left[\mathbf{p} - q_i\mathbf{A}(\mathbf{r}_i)\right]^2}{2m_i} + \sum_{i>j} \frac{q_i q_j}{4\pi\epsilon_0|\mathbf{r}_i - \mathbf{r}_j|} + \frac{\epsilon_0}{2}\int_V dV \left[\mathbf{E}_\perp^2 + c^2\mathbf{B}^2\right], \quad (2.10)$$

where the first term accounts for the kinetic energy of the charges and the light–matter coupling, the second term corresponds to the usual Coulomb instantaneous interaction, and the third term is the electromagnetic energy of the system. This Hamiltonian is known as the *minimal coupling Hamiltonian*. Note that despite it being expressed explicitly in terms of the electric and magnetic fields, these are functionals of the canonical coordinates and momenta.

Before proceeding to quantize the Hamiltonian, note that, in general, we can write the vector potential as an expansion in reciprocal space given by

$$\mathbf{A}(\mathbf{r}, t) = \frac{1}{(2\pi)^{3/2}}\int d^3k \sum_{\lambda=1,2} \left[A_\lambda(\mathbf{k}, t)\mathbf{e}_\lambda e^{i\mathbf{k}\cdot\mathbf{r}} + \text{c.c.}\right] \qquad (2.11)$$

where $\mathbf{e}_\lambda$ are unitary orthogonal vectors representing the two only possible directions of the purely transversal field. The functions $A_\lambda(\mathbf{k}, t)$ are determined by replacing this expansion into Eq. (2.7).[1] We can also express the canonical momenta in reciprocal space, which allow us to rewrite the electromagnetic energy of the system as:

---

[1] This expression is valid both in the presence and absence of sources. It is worth mentioning that in the case of the free field it is possible to write each term as $A_{\mathbf{k},\lambda}(\mathbf{r}, t) = A_{\mathbf{k},\lambda}(\mathbf{r})e^{-i\omega(\mathbf{k})t}$, where the $A_{\mathbf{k},\lambda}(\mathbf{r})$ satisfy the homogeneous Helmholtz equation $(\nabla^2 + k^2)A_{\mathbf{k},\lambda}(\mathbf{r}) = 0$. In this case the explicit time dependence of $\mathcal{H}_{\text{EM}}$ disappears, as the energy is conserved.

$$\mathcal{H}_{\text{EM}} = \frac{\epsilon_0}{2} \int d^3 \mathbf{k} \sum_{\lambda=1,2} \left[\frac{\Pi_\lambda^2(\mathbf{k}, t)}{\epsilon_0^2} + c^2 k^2 A_k^2(\mathbf{k}, t) \right]. \tag{2.12}$$

Note that the explicit dependence with time of Eq. (2.12) means that the energy of the electromagnetic field alone is not conserved, but rather the energy of the full coupled system.

Now, we are ready to quantize the Hamiltonian in Eq. (2.10) by transforming the coordinates and momenta into operators. The standard procedure to find the quantum description from a classical theory is known as canonical quantization. This method was first introduced in 1926 by Paul Dirac in his PhD thesis [3, 4] and consists on imposing quantum commutation relations to the canonical Poisson brackets, i.e.,

$$\{r_{i\alpha}, p_{j\beta}\} = \delta_{ij}\delta_{\alpha\beta} \rightarrow \left[\hat{r}_{i\alpha}, \hat{p}_{j\beta} \right] = i\hbar\delta_{ij}\delta_{\alpha\beta}, \tag{2.13a}$$

$$\{A_\lambda(\mathbf{k}, t), \Pi_{\lambda'}(\mathbf{k}', t)\} = \delta_{\lambda\lambda'}\delta(\mathbf{k} - \mathbf{k}') \rightarrow \left[\hat{A}_\lambda(\mathbf{k}, t), \hat{\Pi}_{\lambda'}(\mathbf{k}', t) \right] = i\hbar\delta_{\lambda\lambda'}\delta(\mathbf{k} - \mathbf{k}'). \tag{2.13b}$$

By replacing these new variables in the form of operators in Eqs. (2.10) and (2.12) we thus find the quantum Hamiltonian of the light–matter system in the Coulomb gauge. Inspection of Eqs. (2.12) and (2.13b) reveals a clear resemblance with the quantum harmonic oscillator. This motivates us to introduce the ladder operators

$$\hat{a}_{\mathbf{k},\lambda}(t) = \sqrt{\frac{\epsilon_0}{2\hbar\omega(\mathbf{k})}} \left[\omega(\mathbf{k})\hat{A}_\lambda(\mathbf{k}, t) + \frac{i}{\epsilon_0}\hat{\Pi}_\lambda(\mathbf{k}, t) \right], \tag{2.14}$$

which satisfy the bosonic commutation relations, i.e., $[\hat{a}_{\mathbf{k},\lambda}, \hat{a}_{\mathbf{k}',\lambda'}] = 0$, $[\hat{a}_{\mathbf{k},\lambda}^\dagger, \hat{a}_{\mathbf{k}',\lambda'}^\dagger] = 0$, and $[\hat{a}_{\mathbf{k},\lambda}, \hat{a}_{\mathbf{k}',\lambda'}^\dagger] = \delta_{\lambda\lambda'}\delta(\mathbf{k} - \mathbf{k}')$. With this definition we can now obtain the vector potential operator and its canonical momentum in reciprocal space as functions of the ladder operators:

$$\hat{A}_\lambda(\mathbf{k}, t) = \sqrt{\frac{\hbar}{2\epsilon_0\omega(\mathbf{k})}} \left(\hat{a}_{\mathbf{k},\lambda}(t) + \hat{a}_{\mathbf{k},\lambda}^\dagger(t) \right), \tag{2.15a}$$

$$\hat{\Pi}_\lambda(\mathbf{k}, t) = -i\sqrt{\frac{\hbar\omega(\mathbf{k})\epsilon_0}{2}} \left(\hat{a}_{\mathbf{k},\lambda}(t) - \hat{a}_{\mathbf{k},\lambda}^\dagger(t) \right), \tag{2.15b}$$

and using Eq. (2.11) and the relations between vector potential and electric and magnetic fields we can obtain the expressions for the field quantum operators in real space:

$$\hat{\mathbf{A}}(\mathbf{r}, t) = \frac{1}{(2\pi)^{3/2}} \int d^3 \mathbf{k} \sum_{\lambda=1,2} \sqrt{\frac{\hbar}{2\epsilon_0\omega(\mathbf{k})}} \left(\hat{a}_{\mathbf{k},\lambda}(t)\mathbf{e}_{\mathbf{k},\lambda}(\mathbf{r}) + \text{H.c.} \right), \tag{2.16a}$$

$$\hat{\mathbf{E}}_\perp(\mathbf{r}, t) = -\frac{i}{(2\pi)^{3/2}} \int d^3\mathbf{k} \sum_{\lambda=1,2} \sqrt{\frac{\hbar\omega(\mathbf{k})}{2\epsilon_0}} \left(\hat{a}_{\mathbf{k},\lambda}(t)\mathbf{e}_{\mathbf{k},\lambda}(\mathbf{r}) - \text{H.c.} \right), \quad (2.16b)$$

$$\hat{\mathbf{B}}(\mathbf{r}, t) = \frac{1}{(2\pi)^{3/2}} \int d^3\mathbf{k} \sum_{\lambda=1,2} \sqrt{\frac{\hbar}{2\epsilon_0\omega(\mathbf{k})}} \left(\hat{a}_{\mathbf{k},\lambda}(t)\left[\nabla \times \mathbf{e}_{\mathbf{k},\lambda}(\mathbf{r})\right] + \text{H.c.} \right),$$

$$(2.16c)$$

where we now have introduced a general spatial dependence on the unitary vectors $\mathbf{e}_{\mathbf{k},\lambda}$. We can finally write the full quantum minimal coupling Hamiltonian using the ladder operators, with now the electromagnetic energy described as a sum of quantum harmonic oscillators[2]:

$$\hat{H} = \sum_i \frac{1}{2m_i}\left[\hat{\mathbf{p}}_i - q_i\hat{\mathbf{A}}(\mathbf{r}_i)\right]^2 + \sum_{i>j} \frac{q_i q_j}{4\pi\epsilon_0|\hat{\mathbf{r}}_i - \hat{\mathbf{r}}_j|} + \sum_{\mathbf{k},\lambda} \hbar\omega(\mathbf{k})\left(\hat{a}_{\mathbf{k}\lambda}^\dagger \hat{a}_{\mathbf{k}\lambda} + \frac{1}{2}\right).$$

$$(2.17)$$

2.1.3 Dipolar Hamiltonian

In the previous section we defined the appropriate quantum operators to describe the quantum fields and the electromagnetic Hamiltonian in terms of the standard creation and annihilation operators. However, we did not explicitly express in those terms the light–matter interaction Hamiltonian, defined as $\hat{H} = \sum_i \frac{1}{2m_i}\left[-2q_i\hat{\mathbf{p}}_i \cdot \hat{\mathbf{A}}(\mathbf{r}_i) + q_i^2\hat{\mathbf{A}}^2(\mathbf{r}_i)\right]$ in the Coulomb gauge. The choice of gauge can impact the physical meaning of many magnitudes, the system modeling, and the numerical accuracy of the description. For instance, in systems where great spatial precision is not required, e.g. in dynamical interactions with a laser, Eq. (2.17) is the most suitable description [1, 5]. However, in most common scenarios in quantum optics the dipole moments of the emitters and the electric field are the most convenient operators. We devote this section to transform the minimal coupling Hamiltonian to describe the light–matter interaction in terms of the electric field.

The following treatment is founded on the original theoretical work of Maria Goeppert-Mayer in 1931 [6], used in the early quantum radiation theory. This method was then generalized by Power and Zienau in 1959 by completing the description when light and matter where treated as a closed dynamical system [7]. Finally, in 1971 Woolley developed a more fundamental view of the transformation [8]. This consists in the unitary transformation of the type $\hat{U} = e^{-i\hat{S}}$ with the generator operator

---

[2]We see that in the vacuum state of the system the energy is $E_0 = \frac{1}{2}\sum_{\mathbf{k},\lambda} \hbar\omega(\mathbf{k})$. The frequencies $\omega(\mathbf{k})$ have no upper bound, so E_0 diverges. However, this is not a problem since expectation values only depend on energy differences and not absolute energies, so the divergence of the vacuum state does not appear in any physical observable.

$$\hat{S} = \frac{1}{\hbar} \int_V d^3 \mathbf{r} \hat{\mathbf{P}}(\mathbf{r}) \cdot \hat{\mathbf{A}}(\mathbf{r}) \tag{2.18}$$

where $\hat{\mathbf{P}}(\mathbf{r})$ is the polarization field operator of the matter, which may be written in terms of the electric multipole moments[3] [9]:

$$\hat{\mathbf{P}}(\mathbf{r}) = \sum_i \hat{\boldsymbol{\mu}}_i \delta(\mathbf{r} - \mathbf{r}_i) - \hat{\mathbf{Q}}_i \nabla \delta(\mathbf{r} - \mathbf{r}_i) + \cdots , \tag{2.19}$$

where $\hat{\boldsymbol{\mu}}_i = \sum_j q_j(\hat{\mathbf{r}}_i - \hat{\mathbf{r}}_j)$ and $\hat{\mathbf{Q}}_i = -1/2 \sum_j q_j(\hat{\mathbf{r}}_i - \hat{\mathbf{r}}_j)(\hat{\mathbf{r}}_i - \hat{\mathbf{r}}_j)$ are respectively the dipole and quadrupole moments of a set of charges q_j with center of mass at $\mathbf{r}_i$, typically corresponding to atoms or molecules. The transformation $\hat{H}_{new} = \hat{U} \hat{H} \hat{U}^\dagger$ changes the Hamiltonian, without modifying the underlying physics, into the so-called *multipolar Hamiltonian*. This is commonly known as the Power–Zienau–Woolley transformation [1, 2]. In the following we present how each term of the Hamiltonian is transformed, for a more detailed explanation see [10]:

$$\hat{U} \left[\sum_i \frac{\left(\hat{\mathbf{p}}_i - q_i \hat{\mathbf{A}}(\mathbf{r}_i)\right)^2}{2m_i} \right] \hat{U}^\dagger = \sum_i \frac{\hat{\mathbf{p}}_i^2}{2m_i} - \int d^3 \mathbf{r} \hat{\mathbf{M}}(\mathbf{r}) \cdot \hat{\mathbf{B}}(\mathbf{r}) \tag{2.20a}$$

$$+ \frac{1}{2} \int d^3 \mathbf{r}' \int d^3 \mathbf{r} \hat{\mathbf{B}}(\mathbf{r}) \hat{\mathbf{O}}(\mathbf{r}, \mathbf{r}') \hat{\mathbf{B}}(\mathbf{r}'),$$

$$\hat{U} \hat{H}_{EM} \hat{U}^\dagger = \hat{H}_{EM} - \frac{1}{\epsilon_0} \int d^3 \mathbf{r} \hat{\mathbf{P}}_\perp(\mathbf{r}) \cdot \hat{\mathbf{D}}_\perp(\mathbf{r}) + \frac{1}{2\epsilon_0} \int d^3 \mathbf{r} \hat{\mathbf{P}}_\perp^2(\mathbf{r}), \tag{2.20b}$$

$$\hat{U} \hat{V}_{\text{Coulomb}} \hat{U}^\dagger = \hat{V}_{\text{Coulomb}}, \tag{2.20c}$$

where $\hat{\mathbf{M}}(\mathbf{r})$ and $\hat{\mathbf{O}}(\mathbf{r}, \mathbf{r}')$ are the magnetization and dimagnetization fields respectively. The corresponding magnetic and diamagnetic terms are in general not important, as their order of magnitude is always smaller than the electric dipole component and usually only become relevant in high external static magnetic fields. For the purpose of this thesis, we now perform the *dipole approximation*,[4] which consists in just considering the dipolar term in Eq. (2.19) and neglecting all higher multipoles. Since the magnetic dipole interaction is of the same order as the electric quadrupole

---

[3]The more fundamental definition for the polarization field is $\hat{\mathbf{P}}(\mathbf{r}) = \sum_{i,j} q_i^{(j)}(\mathbf{r}_i - \mathbf{r}_j) \int_0^1 ds \delta^3 \left[\mathbf{r} - \mathbf{r}_j - s(\mathbf{r}_i - \mathbf{r}_j)\right]$. This is more cumbersome and less intuitive, so we instead present $\hat{\mathbf{P}}(\mathbf{r})$ directly as a multipole expansion. The connection between each expression can be found in [2].

[4]This approximation is completely equivalent to the *long-wavelength approximation*, in which the charges conforming each dipole are very close compared to the EM wavelength and thus experience the same fields, i.e. for $\boldsymbol{\mu} = q_i \mathbf{r}_i + q_j \mathbf{r}_j$ the fields satisfy $\mathbf{A}(\mathbf{r}_i) \approx \mathbf{A}(\mathbf{r}_j)$. This is analogous to neglect the effects of higher electric multipoles, as they are more significant as the spatial structure of the collection of charges becomes important.

one [2], this approximation directly eliminates all magnetic interactions, leading to the *dipolar Hamiltonian*:

$$\hat{H} = \sum_i \frac{\hat{\mathbf{p}}_i^2}{2m_i} + \sum_{i>j} \frac{q_i q_j}{4\pi\epsilon_0 |\hat{\mathbf{r}}_i - \hat{\mathbf{r}}_j|} - \frac{1}{\epsilon_0} \sum_\alpha \hat{\boldsymbol{\mu}}_\alpha \cdot \hat{\mathbf{D}}_\perp(\mathbf{r}) + \frac{1}{2\epsilon_0} \int d^3 r \hat{\mathbf{P}}_\perp^2(\mathbf{r}) + \sum_{\mathbf{k},\lambda} \hbar\omega(\mathbf{k}) \hat{a}_{\mathbf{k}\lambda}^\dagger \hat{a}_{\mathbf{k}\lambda}, \quad (2.21)$$

where now $\hat{\mathbf{P}}(\mathbf{r}) \approx \sum_\alpha \hat{\boldsymbol{\mu}}_\alpha \delta(\mathbf{r} - \mathbf{r}_\alpha)$ in the dipole approximation. Note that the indices i, j represent individual charges, while the index α indicates a single dipole of a collection of charges. This distinction is particularly relevant in this thesis, where we will treat collections of molecules that present an electric dipole moment that depends on the internal degrees of freedom of the molecule.

The Hamiltonian of Eq. (2.21) represents the starting point of most theoretical descriptions in this thesis. However, in most practical cases direct use of this Hamiltonian is very cumbersome and some simplifications are needed. In the next section we focus on the material part of the Hamiltonian in order to achieve a more convenient description for our purposes. Later, in Sect. 2.3, we will focus our attention on a different form of the light–matter Hamiltonian that is more appropriate for CQED.

2.2 Molecular Hamiltonian

In this section we will focus on the material part of Eq. (2.21):

$$\hat{H}_{\text{mat}} = \sum_i \frac{\hat{\mathbf{p}}_i^2}{2m_i} + \sum_{i>j} \frac{q_i q_j}{4\pi\epsilon_0 |\hat{\mathbf{r}}_i - \hat{\mathbf{r}}_j|}. \quad (2.22)$$

As discussed above, the nature of this Coulomb interaction between charges is instantaneous, and deals exclusively with the longitudinal part of the electric field, $\mathbf{E}_\parallel(\mathbf{r})$. While Eq. (2.22) is completely general, in this thesis we do not deal with interacting unbound charges, but rather with interacting bound systems of particles. For the purposes of this thesis, this will represent interacting molecules, but it also may describe atoms, quantum dots, nanoparticles, etc.

Let us first rewrite the Hamiltonian for this many-body problem in the following way:

$$\hat{H}_{\text{mat}} = \sum_i \hat{H}_{\text{mol}}^{(i)} + \sum_{i>j} \hat{H}_{\text{int}}^{(ij)}, \quad (2.23)$$

where $\hat{H}_{\text{mol}}^{(i)}$ describes the Hamiltonian of a single molecule, and $\hat{H}_{\text{int}}^{(ij)}$ the interaction term between the i-th and j-th molecules.[5] In this section we discuss these two Hamiltonians for the case of neutral organic molecules.

---

[5]Note that, depending on the context, we use the indices i and j to represent either individual charged particles or molecules.

2.2.1 *Born–Oppenheimer Approximation*

Let us focus first on the description of a single molecule, for which the Hamiltonian in Eq. (2.22) is still valid. We can rewrite it in a more convenient way, by separating explicitly the N_e electrons and the N_n nuclei conforming the molecule:

$$\hat{H}_{\text{mol}} = \sum_{i=1}^{N_e} \frac{\hat{\mathbf{p}}_i^2}{2m_e} + \sum_{j=1}^{N_n} \frac{\hat{\mathbf{P}}_j^2}{2M_j} + \hat{V}_{\text{ee}}(\hat{\mathbf{r}}_i) + \hat{V}_{\text{nn}}(\hat{\mathbf{R}}_i) + \hat{V}_{\text{en}}(\hat{\mathbf{r}}_i, \hat{\mathbf{R}}_j), \qquad (2.24)$$

where the interaction potentials are:

$$\hat{V}_{\text{ee}}(\hat{\mathbf{r}}_i) = \frac{e^2}{4\pi\epsilon_0} \sum_{i=1}^{N_e} \sum_{j>i}^{N_e} \frac{1}{|\hat{\mathbf{r}}_i - \hat{\mathbf{r}}_j|}, \qquad (2.25a)$$

$$\hat{V}_{\text{nn}}(\hat{\mathbf{R}}_i) = \frac{e^2}{4\pi\epsilon_0} \sum_{i=1}^{N_n} \sum_{j>i}^{N_n} \frac{Z_i Z_j}{|\hat{\mathbf{R}}_i - \hat{\mathbf{R}}_j|}, \qquad (2.25b)$$

$$\hat{V}_{\text{en}}(\hat{\mathbf{r}}_i, \hat{\mathbf{R}}_j) = -\frac{e^2}{4\pi\epsilon_0} \sum_{i=1}^{N_e} \sum_{j=1}^{N_n} \frac{Z_j}{|\hat{\mathbf{r}}_i - \hat{\mathbf{R}}_j|}, \qquad (2.25c)$$

where the $\hat{\mathbf{r}}_i$ and $\hat{\mathbf{R}}_j$ with an explicit subindex outside a sum represent the dependence with the coordinates of all charges. Note that each nucleus can have different masses M_i and charges Z_i, being this completely general for any molecule.

2.2.1.1 Adiabatic Representation

Typically, directly computing the energies and wavefunctions associated with the Hamiltonian in Eq. (2.24) without approximations is a virtually impossible task for typical organic molecules. For example, the anthracene molecule has 24 nuclei and 94 electrons, each of them having 3 spatial degrees of freedom, making a total of 354 different variables in the time independent Schrödinger equation. In the following we will discuss the *Born–Oppenheimer approximation* (BOA), which allows to describe the molecule in a less complicated manner. This approximation was first proposed in 1927 by Max Born and J. Robert Oppenheimer [11, 12] and it is still commonly used in quantum chemistry today. In order to present it, let us first gather terms in Eq. (2.24) and rewrite it as

$$\hat{H}_{\text{mol}} = \sum_{j=1}^{N_n} \frac{\hat{\mathbf{P}}_j^2}{2M_j} + \hat{H}_e(\hat{\mathbf{r}}_i; \mathbf{R}_j), \qquad (2.26)$$

where the electronic Hamiltonian $\hat{H}_e$ contains by definition all electronic contributions and all the nuclear interactions. Since we have separated the nuclear kinetic energy $\hat{T}_n = \sum_{j=1}^{N_n} \frac{\hat{P}_j^2}{2M_j}$, the electronic Hamiltonian only depends parametrically on the nuclear degrees of freedom. Diagonalization of $\hat{H}_e$ yields a set of electronic eigenstates $\{\Phi_k(\mathbf{R}_j)\}$ that inherit this parametric dependence and that satisfy

$$\hat{H}_e(\hat{\mathbf{r}}_i; \mathbf{R}_j)\Phi_k(\mathbf{r}_i; \mathbf{R}_j) = V_k(\mathbf{R}_j)\Phi_k(\mathbf{r}_i; \mathbf{R}_j), \tag{2.27}$$

where the $V_k(\mathbf{R}_j)$ are the so-called electronic *potential energy surfaces* (PESs). The adiabatic wavefunctions $\{\Phi_k(\mathbf{R}_j)\}$ constitute a complete and orthonormal set. It can therefore be used as a basis set in which to expand the total wavefunction of the system

$$\Psi(\mathbf{r}_i, \mathbf{R}_j) = \sum_k \chi_k(\mathbf{R}_j)\Phi_k(\mathbf{r}_i; \mathbf{R}_j), \tag{2.28}$$

where the nuclear wavefunctions $\chi_k(\mathbf{R}_j)$ act as expansion coefficients. This is known as the Born–Huang expansion [13] and it is formally exact when an infinite number of electronic states are included. Inserting this into the full Schrödinger equation of the system $\hat{H}_{\text{mol}}\Psi(\mathbf{r}_i, \mathbf{R}_j) = E\Psi(\mathbf{r}_i, \mathbf{R}_j)$ leads to the set of coupled differential equations

$$\left[\hat{T}_n + V_k(\mathbf{R}_j)\right]\chi_k(\mathbf{R}_j) + \sum_{k'}\hat{\Lambda}_{kk'}(\mathbf{R}_j)\chi_{k'}(\mathbf{R}_j) = E\chi_k(\mathbf{R}_j), \tag{2.29}$$

where the operator $\hat{\Lambda}_{kk'}(\mathbf{R}_j) = \langle\Phi_k(\mathbf{r}_i; \mathbf{R}_j)|\hat{T}_n|\Phi_{k'}(\mathbf{r}_i; \mathbf{R}_j)\rangle_{\mathbf{r}_i} - \hat{T}_n\delta_{kk'}$ accounts for *nonadiabatic* couplings between the different PESs, describing the dynamical interaction between electronic and nuclear motion. The subindex $\mathbf{r}_i$ represents integration over all the electronic degrees of freedom.

The Born–Oppenheimer approximation consists now in describing the full wavefunction of the system by the ansatz

$$\Psi(\mathbf{r}_i, \mathbf{R}_j) = \chi_k(\mathbf{R}_j)\Phi_k(\mathbf{r}_i; \mathbf{R}_j), \tag{2.30}$$

i.e., any molecular state is represented by a single product of an adiabatic electronic state and a nuclear wavefunction. By replacing this in the Schrödinger equation we get a similar expression to Eq. (2.29) that results in

$$\left[\hat{T}_n + V_k(\mathbf{R}_j) - \hat{\Lambda}_{kk}(\mathbf{R}_j)\right]\chi_k(\mathbf{R}_j) = E\chi_k(\mathbf{R}_j), \tag{2.31}$$

where $\hat{\Lambda}(\mathbf{R}_j)$ is now a purely diagonal operator. Although the approximation effectively decouples the nuclear and electronic degrees of freedom, calculating $\hat{\Lambda}(\mathbf{R}_j)$ is not generally an easy task. For the sake of clarity let us now consider that all the nuclei have the same mass $M = M_j$. Then we can rewrite the expression for $\hat{\Lambda}_{kk'}(\mathbf{R}_j)$ as

$$\hat{\Lambda}_{kk'}(\mathbf{R}_j) = \frac{1}{2M}\left(2\hat{\mathbf{F}}_{kk'}(\mathbf{R}_j)\cdot\nabla + \hat{G}_{kk'}(\mathbf{R}_j)\right), \tag{2.32}$$

where $\hat{\mathbf{F}}_{kk'}(\mathbf{R}_j) = \langle\Phi_k(\mathbf{R}_j)|\nabla\Phi_{k'}(\mathbf{R}_j)\rangle$ and $\hat{G}_{kk'}(\mathbf{R}_j) = \langle\Phi_k(\mathbf{R}_j)|\nabla^2\Phi_{k'}(\mathbf{R}_j)\rangle$ are the derivative coupling vector and the scalar coupling respectively. We can now argue doing an additional adiabatic approximation by neglecting the term $\Lambda_{kk}(\mathbf{R}_j)$ in Eq. (2.31). The separability of nuclear and electronic wavefunctions in BOA rests on the large mass difference between nuclei and electrons. This is reflected in Eq. (2.32), where it is clear that a large nuclear mass leads to small nonadiabatic couplings. Therefore, this nonadiabatic term is typically ignored in most calculations, as it is often too complicated to calculate in organic molecules and represents a negligible correction to the electronic PESs. Note that this approximation is often known as the *adiabatic approximation*. However, the ansatz of Eq. (2.30) is also often called Born–Huang approximation while neglecting the diagonal nonadiabatic term is the BOA [14]. Every reference to the BOA throughout this thesis will be referring to the latter denotation.

Effectively, the BOA considers that the electrons instantaneously adapt to the nuclear motion so that the energy updates instantly when changing the configuration of the nuclei. This however only applies for an isolated single electronic state and can break down when two states come close. This is expressed in the dependence of the nonadiabatic coupling with $\hat{\mathbf{F}}_{kk'}(\mathbf{R}_j)$, which by using the Hellmann–Feynman theorem can be rewritten as

$$\hat{\mathbf{F}}_{kk'}(\mathbf{R}_j) = \frac{\langle\Phi_k(\mathbf{R}_j)|\nabla\hat{H}_e|\Phi_{k'}(\mathbf{R}_j)\rangle}{V_{k'}(\mathbf{R}_j) - V_k(\mathbf{R}_j)}. \tag{2.33}$$

It becomes immediately apparent that the nonadiabatic coupling will increase when the electronic PESs come close to each other, and even diverge if the energies are equal. These points of degeneracy are known as conical intersections, and they have a central role in nonadiabatic transitions [15].

2.2.1.2 Diabatic Representation

While the adiabatic representation is useful for most calculations in organic molecules, it is still difficult to solve when the nonadiabatic coupling vector $\hat{\mathbf{F}}_{kk'}(\mathbf{R}_j)$ is relevant. This nonlocal operator, that describes coupling between electronic states through nuclear motion, is not an intuitive quantity to work with, and can even present a singular behavior in the vicinity of intersection between PESs. For these situations a *diabatic basis* is more favorable. This is achieved through a unitary transformation $\hat{D}$ of the electronic basis set $\tilde{\Phi}(\mathbf{R}_j) = \hat{D}(\mathbf{R}_j)\Phi(\mathbf{R}_j)$. In general it can be shown [14] that the equations Eq. (2.29) now read

$$\left[-\frac{1}{2M}\left(\nabla + \tilde{\mathbf{F}}(\mathbf{R}_j)\right) + \tilde{V}(\mathbf{R}_j)\right]\chi(\mathbf{R}_j) = E\chi(\mathbf{R}_j). \tag{2.34}$$

The diagonal PES matrix has transformed through $\tilde{V}(\mathbf{R}_j) = \hat{D}^\dagger(\mathbf{R}_j)\hat{V}(\mathbf{R}_j)\hat{D}(\mathbf{R}_j)$ and the transformed nonadiabatic derivative coupling is

$$\tilde{\mathbf{F}}(\mathbf{R}_j) = \hat{D}^\dagger(\mathbf{R}_j)\hat{\mathbf{F}}(\mathbf{R}_j)\hat{D}(\mathbf{R}_j) + \hat{D}^\dagger(\mathbf{R}_j)\left(\nabla\hat{D}(\mathbf{R}_j)\right). \tag{2.35}$$

Note that this is general for any kind of unitary transformation to an arbitrary basis. We can now fix the necessary conditions to transform to the diabatic basis, i.e., one in which the nonadiabatic couplings in Eq. (2.34) are eliminated. This can be achieved if the transformation matrix $\hat{D}$ satisfies the following condition

$$\hat{\mathbf{F}}(\mathbf{R}_j)\hat{D}(\mathbf{R}_j) + \left(\nabla\hat{D}(\mathbf{R}_j)\right) = 0. \tag{2.36}$$

We can see that for such a transformation $\tilde{\mathbf{F}}(\mathbf{R}_j) = 0$ and the Schrödinger equation in the diabatic basis will thus read

$$\left[-\frac{1}{2M}\nabla + \tilde{V}(\mathbf{R}_j)\right]\chi(\mathbf{R}_j) = E\chi(\mathbf{R}_j). \tag{2.37}$$

In this basis the diabatic PESs significantly change with respect to the adiabatic picture, and are coupled through the offdiagonal terms $\tilde{V}_{kk'}(\mathbf{R}_j)$, which are much easier to compute.

One relevant question is whether a pure diabatic basis exists, i.e., does Eq. (2.36) have in general a solution. The answer is that strictly diabatic states are only possible in a one-dimensional problem such as a diatomic molecule [16]. The efforts to find the exact diabatic transformation have thus turned to finding the matrix $\hat{D}$ such that $\tilde{\mathbf{F}}(\mathbf{R}_j)$ is not exactly zero, but negligible. The basis set is thus formed by the so-called *quasi-diabatic states*, essential for many numerical simulations.

2.2.2 Intermolecular Forces

We now focus on the second part of Eq. (2.23) and calculate the interaction energy for two different charged systems A and B as schematized in Fig. 2.1. For this it is particularly useful to describe the systems as two separated charge distributions $\rho_A(\mathbf{r})$ and $\rho_B(\mathbf{r})$ defined as $\rho_\alpha(\mathbf{r}) = \sum_i^{N_\alpha} q_i\delta(\mathbf{r} - \mathbf{r}_i)$. The interaction energy reads

$$V_{AB} = \frac{1}{4\pi\epsilon_0}\int d^3\mathbf{r}\int d^3\mathbf{r}'\frac{\rho_A(\mathbf{r})\rho_B(\mathbf{r}')}{|\mathbf{r} - \mathbf{r}'|}. \tag{2.38}$$

If the separation between the center of masses of each distribution, $|\hat{\mathbf{R}}| = |\hat{\mathbf{r}}_A - \hat{\mathbf{r}}_B|$ is much larger than the sizes of each distribution, we may expand V_{AB} in a multipole series with respect to $\mathbf{r}_A$ and $\mathbf{r}_B$, resulting in:

Fig. 2.1 Scheme of two
interacting charge
distributions in space

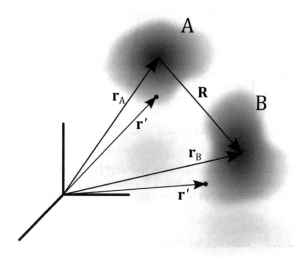

$$\hat{V}_{AB}(\mathbf{R}) = \frac{1}{4\pi\epsilon_0}\left[\frac{q_A q_B}{|\mathbf{R}|} + \frac{q_A \hat{\boldsymbol{\mu}}_B \cdot \mathbf{e_R}}{|\mathbf{R}|^2} - \frac{q_B \hat{\boldsymbol{\mu}}_A \cdot \mathbf{e_R}}{|\mathbf{R}|^2}\right.$$
$$\left. + \frac{\hat{\boldsymbol{\mu}}_A \cdot \hat{\boldsymbol{\mu}}_B - 3(\hat{\boldsymbol{\mu}}_A \cdot \mathbf{e_R})(\hat{\boldsymbol{\mu}}_B \cdot \mathbf{e_R})}{|\mathbf{R}|^3} + \cdots\right], \quad (2.39)$$

where $\mathbf{e_R} = \mathbf{R}/|\mathbf{R}|$ is the unitary vector connecting the two systems, q_α is the total
charge of each distribution, and $\hat{\boldsymbol{\mu}}_\alpha = \int d^3\mathbf{r}\rho_\alpha(\hat{\mathbf{r}})(\hat{\mathbf{r}} - \hat{\mathbf{r}}_\alpha)$ is the total dipole of each
system.

In the following we only consider neutral molecules as our charge distributions
and we ignore higher order multipoles in the expansion, as their contribution to the
interaction is negligible. Therefore, now the total interaction Hamiltonian between
molecules is purely dipolar:

$$\hat{H}_{dd} = \frac{1}{4\pi\epsilon_0}\sum_{i>j}\frac{\hat{\boldsymbol{\mu}}_i \cdot \hat{\boldsymbol{\mu}}_j - 3(\hat{\boldsymbol{\mu}}_i \cdot \mathbf{e}_{\mathbf{R}_{ij}})(\hat{\boldsymbol{\mu}}_j \cdot \mathbf{e}_{\mathbf{R}_{ij}})}{|\mathbf{R}_{ij}|^3}. \quad (2.40)$$

This interaction gives rise to the well-known van der Waals forces and to Förster resonance energy transfer (FRET) between molecules [17–19]. Finally, it is important
to emphasize that this term only accounts for the instantaneous Coulomb interaction,
and that it is mediated by the longitudinal electric near field. For much larger distances between molecules, a retarded interaction comes into play,[6] and its description
requires the full light–matter Hamiltonian.

---

[6]This effect emerges because the speed of light is finite. When the information of a particular charge
configuration reaches an emitter situated far away, these charges have already rearranged, so that
the emitter response is no longer in phase. This arises for distances much larger than the wavelength
corresponding to the characteristic absorption frequency of the emitters [20]. In the systems that

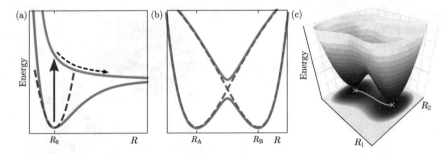

Fig. 2.2 Some common conceptual energy landscapes in chemistry. **a** Bond dissociation picture. **b** Isomerization scenario between two stable configurations R_A and R_B. **c** Two-dimensional ground state surface with two stable configurations connected by a minimum energy path (light gray line)

2.2.3 Chemical Processes

In here we discuss the relevance of the Born–Oppenheimer approximation in organic molecules and the wide range of chemical processes these compounds present. The potential energy surfaces of an organic molecule, or a particular reaction, contain essential chemical information of the system. However, finding the full energy landscape of molecules and reactions is an arduous task due to the great number of DoF involved, resulting in intricate multi-dimensional hypersurfaces. Furthermore, most of the nuclear configurations of a molecule do not play a significant role in any molecular process, so often calculating the full PES landscape is a superfluous effort. Usually the best strategy is to estimate beforehand which are the relevant configurations of the molecule. In the following we discuss the relevance of some points of the PES in chemical processes, as well as introduce some tools that can be used to obtain chemical information of the system.

The most relevant points are usually the stationary points of the surfaces. These are characterized by the condition $\nabla V(\mathbf{R}) = 0$, and can represent local minima or saddle points of the PES $V(\mathbf{R})$. The minima describe the *equilibrium configurations* of the molecule. Around these points usually a harmonic approximation of the surface is performed, as depicted in Fig. 2.2a. Here, the ground state PES (blue line) is described by a Morse potential, $V_g(R) = D_e(1 - e^{A(R-R_0)})^2$, however close to the equilibrium configuration at R_0 it can be approximated by the corresponding quadratic potential (dashed dark blue line). In this same plot we represent an excited state (orange line) without a definite minimum. In this conceptual scenario, after photoabsorption, the molecule would be promoted to this excited state at R_0, thus experiencing a force $-\partial_R V_e(R)|_{R_0}$ towards a larger R. This is a typical situation in the event of bond dissociation, where R may represent the bond distance between two nuclei, which can dissociate after absorption of a photon.

---

we are concerned with in this thesis, these wavelengths are of the order of hundreds of nanometers, so we can completely disregard retardation effects when dealing with dipole–dipole interactions.

Another example for a stationary configuration is the saddle point. This often represents the *transition state* configuration in a chemical reaction. Consider for instance the scenario depicted in the ground-state PES of Fig. 2.2b, where two equivalent equilibrium configurations R_A and R_B are separated by a local maximum.[7] This may describe the reaction between two isomers of the same molecule, i.e., two different nuclear configurations for the same chemical composition. In the energy landscape of Fig. 2.2b it is also possible to photoexcite the molecule to trigger a photoisomerization reaction. After absorption, the molecule is in the excited state at R_A, from where it will evolve towards the transition state, where often the nonadiabatic coupling becomes important, inducing a *nonradiative transition* to the ground-state PES. These events often happen close to or through conical intersections, as presented in Sect. 2.2. In nonradiative transitions the molecule evolves roughly following the diabatic surfaces, represented as dashed lines in Fig. 2.2b. In Chap. 5 we study an example of a photochemical reaction where diabatic surfaces become certainly useful to study dynamics after photoexcitation.

In the two previous examples we restrict the picture to one-dimensional PESs, but as already mentioned, often this is not the case in real molecules. The PES is here used as a conceptual tool that aids to illustrate the analysis of molecular phenomena. Nevertheless, we can often reduce the dimensionality of the system PES by finding the appropriate combination of coordinates that describe a particular process, thus generating an effective PES. See for example the case in Fig. 2.2c, where a two-dimensional ground-state PES displays two different minima. We can find the *minimum energy path* (MEP) between the two (see its projection in light gray). This combination of R_1, R_2 is called the *reaction coordinate* or *reaction path*. A PES like the ground state in Fig. 2.2b may be the effective PES of a two-dimensional scenario, now being R the corresponding reaction coordinate.

Much effort is devoted to develop methods to compute the MEP that can effectively describe a reaction in a multi-dimensional PES [21]. These methods typically require knowledge of the initial and final configurations of a process, which can be found with minimization routines such as steepest descent by introducing an initial estimation as input. This simplification of the PES often can be used to directly calculate the system dynamics [22]. However, it should be noted than in many complex systems, simulating the actual quantum dynamics of the system can become a rather crude approximation. If the nuclei move very slowly on the PES they behave approximately classically, and then the MEP can be used to compute classical dynamics of the reaction. Many methods use simplified PES to simulate classical dynamics in particular processes (e.g., the widely-used surface hopping method [23, 24]). However, often finding the MEP gives information of intermediate states such as transition states, or metastable configurations. For the sake of illustration, consider that the two-dimensional PES in Fig. 2.2c represents the ground-state surface of a molecule. It is possible to obtain the information about the two minima configurations via minimization routines and then compute the minimum energy path that connects them.

---

[7]Note that while in one dimension the transition state is a local maximum, in general multi-dimensional scenarios this is a saddle point.

With this approach, the transition state between the two is apparent, which allows to compute the energy barrier that separates the two equilibrium configurations.

Energy barriers are central in thermally-driven ground-state chemistry, as it is one of the magnitudes that govern the reaction rate from one chemical species to another. In *transition state theory* (TST) the variance of the rate k of a chemical reaction with temperature depends on the energy barrier E_b as described by the Eyring equation (often known as Eyring–Polanyi equation) [25]

$$k = \kappa \frac{k_B T}{h} e^{-\frac{E_b}{k_B T}}, \tag{2.41}$$

where κ is the transmission coefficient, typically considered equal to one if nonadiabatic effects are negligible close to the energy barrier. This equation resembles the well-known empirical Arrhenius equation [26]. However, it should be noted that the Eyring equation is derived from statistical mechanical arguments. One of the basic ideas in the development of TST and the Eyring equation is that reaction rates can be studied by examining the process close to the transition and equilibrium states. This arises from assuming quasi-equilibrium between reactants and products, where both states reach a Boltzmann thermal distribution of energies. This fails in short-lived states where the time to achieve thermal equilibrium is greater than the lifetime of the state. In these situations TST only gives a rough estimate of the reaction rate.

It should be noted that the Eyring equation arises from classical considerations, and a rigorous quantum rate theory is required when purely quantum effects become relevant, such as nonadiabatic couplings, zero-point energy, and tunneling [27]. In the following we will briefly review an approach based on the correlation function formalism [28–30] that is used in Chap. 6. This states that the rate of a molecular reaction is given by

$$k(T) = \frac{1}{Q_r(T)} \int_0^{t_f \to \infty} C_{ff}(t) dt, \tag{2.42}$$

where $Q_r(T) = \mathrm{tr}[\exp(-\beta \hat{H})]$, with $\beta^{-1} = k_B T$, is the time-dependent partition function of the reactants at temperature T and $C_{ff}(t)$ is the flux-flux autocorrelation function, defined as

$$C_{ff}(t) = \mathrm{tr}[\bar{F} \hat{U}^\dagger(t_c) \bar{F} \hat{U}(t_c)]. \tag{2.43}$$

This correlation function is computed as the trace of a product of operators, where $U(t_c) = \exp(-i \hat{H} t_c)$, with $t_c = t - i\beta/2$, is the complex time evolution operator and $\bar{F}$ represents the symmetrized flux operator

$$\bar{F} = \frac{1}{2M} \left(\hat{\mathbf{P}} \cdot \frac{\partial s(\mathbf{R})}{\partial \mathbf{R}} \delta(s) + \delta(s) \hat{\mathbf{P}} \cdot \frac{\partial s(\mathbf{R})}{\partial \mathbf{R}} \right). \tag{2.44}$$

Here, $\hat{\mathbf{P}}$ is the nuclear momentum operator and the surface dividing the reactant and product states is defined by the zeros of the function $s = s(\mathbf{R})$, e.g., the function $s(R) = R$ corresponds to a dividing surface at $R = 0$. The flux-flux autocorrelation function describes the temporal flux of positive-momenta probability through the dividing surface of a thermally averaged initial state (which is accounted for by the thermal part of the $\hat{U}(t_c)$ operator). Negative values of $C_{ff}(t)$ indicate recrossing of the dividing surface in the opposite direction, thus contributing to a rate decrease. Note that this effectively involves generating the full quantum dynamics up to t_f, as indicated in Eq. (2.42). Nevertheless, relevant purely quantum phenomena, such as tunneling, take place mostly between times 0 and $\hbar\beta$, which corresponds to $t_f \approx$ 27 fs for room temperature. Moreover, it is often useful to treat secondary DoF approximately as, e.g. a thermal bath. In such a scenario it is possible to integrate the flux-flux autocorrelation function up to the typical dissipation time of the system, granted that $C_{ff}(t_f) \approx 0$.

2.2.4 Response to the Electromagnetic Field

The various effects of the electromagnetic field on organic molecules are crucial to the work developed in this thesis, not only in order to comprehend polariton formation but, as discussed above, because some chemical reactions can be triggered by light absorption or by the presence of external electromagnetic fields. Let us thus present and discuss in the following some molecular properties related to the coupling to the electromagnetic field. In the dipolar gauge, the full interaction Hamiltonian is

$$\hat{H}_{int} = -\frac{1}{\epsilon_0} \sum_\alpha \hat{\mu}_\alpha \cdot \hat{\mathbf{D}}_\perp(\mathbf{r}) + \frac{1}{2\epsilon_0} \int d^3\mathbf{r} \hat{\mathbf{P}}_\perp^2(\mathbf{r}). \qquad (2.45)$$

We see that in these expression the only EM contribution comes from the displacement field $\hat{\mathbf{D}}_\perp(\mathbf{r})$, and therefore the second term contributes to the matter Hamiltonian by renormalizing its energy.[8] Here we only discuss the effect on the molecules of the dipolar term $\propto \hat{\mu} \cdot \hat{\mathbf{E}}_\perp(\mathbf{r})$, which dominates the light–matter interaction. The dipole moment operator of the molecules becomes of great relevance in this picture. In general, this operator is defined as

$$\hat{\mu} = -e \sum_i^{N_e} \hat{\mathbf{r}}_i + e \sum_j^{N_n} Z_j \hat{\mathbf{R}}_j \qquad (2.46)$$

for a molecule with center of mass at $\mathbf{R} = 0$. The dipole operator in Eq. (2.46) is expressed in an spatial basis. However, depending on the particular problem, we might be interested in transforming this operator to an adiabatic (BOA) or diabatic basis.

---

[8]In Sect. 2.3 we analyze the relevance of this term when not all EM modes are explicitly considered.

2.2.4.1   Molecular Polarizability

When a molecule[9] is exposed to an external electric field the charges it is composed of tend to rearrange in order to minimize the energy. This can alter the properties of the system, being the electric dipole moment of particular interest. The total dipole moment in the presence of the electric field can be written in terms of *polarizabilities* as

$$\mu = \mu_0 + \alpha \mathbf{E} + \frac{1}{2}\beta \mathbf{E}^2 + \cdots . \tag{2.47}$$

The term μ_0 corresponds to the actual permanent dipole moment of the molecule. The additional terms are collectively known as the *induced dipole moment*, and are characterized by the polarizability α and hyperpolarizabilities of increasingly higher order (such as β). These are tensors that present larger ranks for increasing polarizability order. In the following we are going to focus on the lowest-order modification of the dipole moment, described by the polarizability α, and neglect any effect of the hyperpolarizabilities.

In many situations the polarizability can be of great importance, such as in molecules with permanent zero dipole, where the main contribution to the response to an electromagnetic field is given by the polarizability. Moreover, electric fields with source in other charge distributions may induce dipoles in these molecules so that they experience an interaction. These interactions are known in general as van der Waals forces, which are typically weak and short-range but often play a fundamental role in many diverse fields such as biology or nanotechnology. The polarizability is also of great importance in macroscopic media, where the relative dielectric constant depends on the polarizability of the atoms or molecules that constitute the material through the Clausius–Mossotti relation [31]. Many physical properties of materials depend on the polarizability, which is central to determine the electronic properties and structure of atoms, molecules, clusters, etc.

In general, the polarizability α is a tensor and depends on the frequency ω of the inducing electric field. A complete derivation of this can be found in [32]. We now present a derivation and a brief discussion of the scalar and static ground-state polarizability. Let us first define the polarizability as $\mu = \alpha \cdot \mathbf{E}$. To express the change in interaction energy $E_{\text{int}} = -\mu \cdot \mathbf{E}$ due to an external electric field, let us consider the infinitesimal work the electric field has to do in order to induce a dipole

$$dW = -\mu \cdot d\mathbf{E} = -(\alpha \cdot \mathbf{E}) \cdot d\mathbf{E}. \tag{2.48}$$

By integrating from 0 to $\mathbf{E}$ we find the total change in energy

$$W = -\frac{1}{2}\mathbf{E} \cdot \alpha \cdot \mathbf{E}. \tag{2.49}$$

---

[9]In this thesis we are mainly interested in molecules, but this occurs for any set of charges such as atoms, quantum dots, nanoparticles, etc.

Let us now find the quantum-mechanical expression for the polarizability by considering the interacting Hamiltonian

$$\hat{H} = \hat{H}_0 + \hat{H}_{\text{int}}, \tag{2.50}$$

where the interaction term $\hat{H}_{\text{int}} = -\hat{\boldsymbol{\mu}} \cdot \mathbf{E}$ is treated perturbatively, considering a small external electric field $\mathbf{E}$. The first-order energy shift will simply be $V_0^{(1)} = -\langle 0|\hat{\boldsymbol{\mu}}|0\rangle \cdot \mathbf{E}$, which is zero if there is no permanent dipole in the ground state. The second-order energy shift due to the perturbative potential is thus

$$V_0^{(2)} = \sum_{i \neq 0} \frac{|\langle i|\hat{\boldsymbol{\mu}}|0\rangle|^2}{V_0 - V_i} \mathbf{E}^2, \tag{2.51}$$

where $V_i = \langle i|\hat{H}_0|i\rangle$. By comparing the last equation to Eq. (2.49) we find the quantum-mechanical expression for the scalar and static ground-state polarizability

$$\alpha_0(\omega = 0) = 2\sum_{i \neq 0} \frac{|\langle i|\hat{\boldsymbol{\mu}}|0\rangle|^2}{V_i - V_0}. \tag{2.52}$$

2.2.4.2 Absorption Spectra

In the previous section we restricted our discussion to a very simple picture of the polarizability. A generalization of this tensor is known as the scattering tensor, which describes the scattering processes of electromagnetic radiation with particles. In molecules, we can distinguish between Rayleigh (elastic) and Raman (inelastic) scatterings. A detailed discussion of scattering in molecules and its connection to the polarizability tensor is out of the scope of this theoretical introduction. Nevertheless, for an in-depth explanation the reader may consult chapter two in the book of Bonin and Kresin [32].

We focus now on the absorptive parts of the scattering process. By using the optical theorem [33] we find that the frequency-dependent absorption cross section can be expressed as

$$\sigma(\omega) = \frac{4\pi\omega}{c} \text{Im}\left[f(\omega)\right], \tag{2.53}$$

where $f(\omega)$ is the scattering amplitude at frequency ω, given by

$$f(\omega) = \sum_k \frac{|\langle \Psi_k|\hat{\boldsymbol{\mu}}|\Psi_0\rangle|^2}{\omega_k - \omega_0 - \omega - i\epsilon_k}, \tag{2.54}$$

where the sum runs over all eigenstates $|\Psi_k\rangle$ of the system, being ω_k their energy and ϵ_k representing the corresponding linewidth.

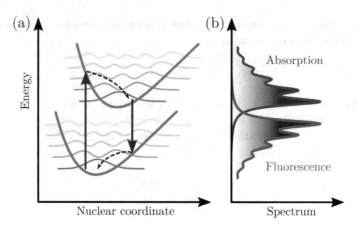

Fig. 2.3 **a** Typical simplified picture of the energy landscape of an organic molecule with different processes represented: absorption (dark red arrow), emission (dark blue arrow), and nuclear relaxation (dashed black arrows). **b** Corresponding absorption and emission (fluorescence) spectra for this model system

Let us review the process of absorption in an organic molecule, schematically illustrated in Fig. 2.3. Initially the molecule is in the electronic ground state, with the corresponding nuclear configuration that minimizes the energy. It can then absorb a photon and be promoted to the first electronic excited state. During this fast transition process the nuclear configuration is unchanged to a *vertical transition*,[10] as Fig. 2.3a shows. This means that the nuclear wavefunction remained unchanged from its ground state shape. However, the equilibrium position in the electronic excited state is often different that the one in the ground state and the nuclear wavepacket is not an eigenstate anymore, and will thus evolve on the new PES. *Kasha's rule* [34] states that molecules will quickly relax nonradiatively towards the lowest energy level, thus changing the configuration of the molecule. Typically this relaxation occurs in much shorter timescales than the lifetime of the excited state. This greatly depends on the molecule, but generally relaxation occurs in tens of femtoseconds up to a picosecond, while excited state lifetimes are of the order of nanoseconds [14, 15, 35, 36]. After this period, emission occurs, also as a vertical transition. Vertical transitions and internal relaxation results in emission of a photon with different frequency than the absorbed photon, i.e. the molecule is *fluorescent*. This can be seen in Fig. 2.3b, where the maximum of the fluorescence (emission) spectrum is different that the absorption one. This difference is the Stokes shift.

The lifetime of the electronic excited state can be calculated using *Fermi's golden rule*. It describes the transition probability from one eigenstate of a discrete system

---

[10]The *Franck–Condon principle* states that the intensity of a vibronic transition (i.e., a change of electronic and vibrational states) is directly proportional to the overlap between the corresponding nuclear wavefunctions. This is based on the assumption of a vertical transition, i.e., that there is some adiabatic separation between nuclear and electronic timescales, similarly as in the Born–Oppenheimer approximation.

to states in a continuum. It was first derived by Paul Dirac [3] using time-dependent perturbation theory, and the resulting transition probability per unit time to first order reads

$$\Gamma_{i \to f} = \frac{2\pi}{\hbar} |\langle f | \hat{H}_{\text{int}} | i \rangle|^2 \rho(E_f), \qquad (2.55)$$

where, for the case of transition to the continuum of EM modes in free-space, $\hat{H}_{\text{int}} = -\hat{\boldsymbol{\mu}} \cdot \hat{\mathbf{E}}_{\perp}$ is the interaction Hamiltonian in the dipole approximation and $\rho(E_f)$ is the density of final states in the continuum. The lifetime of a vibronic state i can be calculated by summing Eq. (2.55) over all possible final states $|f\rangle$. In the case of isolated molecules in free space, this lifetime is often the main contribution to the linewidths ϵ_k of the absorptive scattering amplitude of Eq. (2.54). However, in more complex systems the molecule unavoidably couples to the environment, leading to incoherent processes of energy loss and quantum decoherence [36, 37]. In this thesis we do not apply any open quantum system formalism but instead we analyze and discuss the effect of losses on the results.

2.3 Cavity Quantum Electrodynamics

The field of CQED has proved that the quantum nature of light can be exploited to dramatically modify the behavior of coupled light–matter systems. The microcavity influences the electromagnetic environment so that the interaction between light and matter can be enhanced, leading to many interesting phenomena. In this section we analyze the impact of the cavity on the dipolar Hamiltonian in Eq. (2.21), and review some of the theoretical descriptions used to study strong light–matter coupling.

2.3.1 Electromagnetic Fields in Cavities

The geometrical and material properties of the cavity have an effect on the behavior of light in different ways. A precise microscopic description would encompass the DoF of all atoms forming the medium, leading to very high computational cost for most cavities.[11] A more feasible approach is to describe the medium macroscopically, including the EM response of the cavity material. Typically this can be done by including the relative permittivity and permeability of the medium, which in general are complex functions of frequency. However, a naive extension of the quantization scheme performed in Sect. 2.1 leads to issues when dielectrics are included. The usual plane wave solutions in the Helmholtz equation for free space are now replaced by damped plane waves, which cannot form the complete set of orthonormal functions

---

[11] We note that theoretical efforts in this regard have been made for small plasmonic nanoparticles using time-dependent density functional theory [38].

required to quantize the field. In macroscopic QED [39, 40] this is solved by using creation and annihilation operators with spatial dependence that satisfy the Langevin noise equation in order to accommodate the losses in the medium [41].

Less involved approaches are possible by not including directly the cavity medium in the quantization scheme, but rather approximate it as perfect reflectors that can modify the boundary conditions of the system. For instance, it may force the system to display EM standing waves, as in Fabry–Perot cavities [42], or it might strongly localize the electric field, as in plasmonic nanocavities [43]. This depends on the properties and nature of the cavity, which generally are encoded in the EM fields. As discussed in Sect. 2.1, the form and behavior of the fields is governed by the Maxwell equations. In the following let us focus on two different cavities and analyze how the Hamiltonian can be altered.

2.3.1.1 CQED in a Planar Cavity

We examine the case of two parallel mirrors of a very large area A situated at $z = 0$ and $z = L$ with no sources, such that the fields satisfy the homogeneous Helmholtz equation. The transverse electric field can be thus written as:

$$\hat{\mathbf{E}}_\perp(z, t) = -i \sum_k \sum_{\lambda=1,2} \sqrt{\frac{\hbar \omega_k}{2\epsilon_0 L A}} \left(\hat{a}_{\mathbf{k},\lambda} \mathbf{e}_{\mathbf{k},\lambda} - \text{H.c.} \right) \sin(k_z z - \omega_k t), \quad (2.56)$$

which describes standing waves, where both the frequency and amplitude can be tuned by modifying the geometry of the cavity. This directly influences the dipolar Hamiltonian, in which the light–matter interaction is governed by the transverse electric field. The set of possible EM states allowed by Eq. (2.56) impacts the density of states in Eq. (2.55), thus altering the decay rate of material excitations. This is the so-called *Purcell effect* and will be discussed in more detail later in this section.

The longitudinal part of the electric field, which describes instantaneous Coulomb-like interactions, is also modified by the presence of the mirrors. The material charge distribution inside the cavity can induce a redistribution of charges in the material of the cavity (e.g. a metallic mirror), which can be easily understood in terms of the method of image charges. This introduces a new interaction that can be written in terms of the charge density of the system:

$$V_{\text{imag}} = \frac{1}{2} \frac{1}{4\pi\epsilon_0} \int d^3\mathbf{r} \int d^3\mathbf{s} \frac{\rho(\mathbf{r})\tilde{\rho}(\mathbf{s})}{|\mathbf{r} - \mathbf{s}|}, \quad (2.57)$$

where $\tilde{\rho}(\mathbf{s})$ corresponds to the charge density of the images, which will depend on the geometry of the cavity. Note the new prefactor $1/2$ that we need to add for interaction between the charge distribution and the *induced* charge distribution. The intuition behind this is that as a charge moves a distance dr towards the mirror boundary, the image charge gets also dr closer, so the actual work dW required to move a distance dr corresponds to a standard charge Coulomb interaction that has moved $2dr$.

In a mirror cavity the induced charge distribution is given by $\tilde{\rho}(\mathbf{s}) = -\int d^3\mathbf{r}'\rho(\mathbf{r}')$ $\delta(\mathbf{s} - \sigma\mathbf{r}')$, where $\sigma\mathbf{r}'$ is the virtual location in the mirror corresponding to the charge at $\mathbf{r}'$. Interestingly, it can be shown [9] that for this particular example of a planar cavity the interaction with the images exactly cancels with the longitudinal polarization density, i.e., $\frac{1}{2\epsilon_0}\int d^3\mathbf{r}\hat{\mathbf{P}}_\parallel^2(\mathbf{r}) + V_{\text{imag}} = 0$, and thus the effect of the cavity is only included in the transversal field of Eq. (2.56). This is not the case for more general cavities, such as metallic geometries that can host plasmonic modes, where both the transversal and longitudinal fields are modified.

2.3.1.2   Light–Matter Hamiltonian in the Quasistatic Limit

In order to study the Hamiltonian in a nanoscale cavity such as systems hosting plasmonic modes, we rewrite the minimal coupling Hamiltonian of Eq. (2.17), explicitly written in terms of the EM fields:

$$\hat{H} = \sum_i \frac{1}{2m_i}\left[\hat{\mathbf{p}}_i - q_i\hat{\mathbf{A}}(\mathbf{r}_i)\right]^2 + \sum_{i>j} \frac{q_iq_j}{4\pi\epsilon_0|\hat{\mathbf{r}}_i - \hat{\mathbf{r}}_j|} + \frac{\epsilon_0}{2}\int dV\left(\hat{\mathbf{E}}_\perp^2 + c^2\hat{\mathbf{B}}^2\right).$$

(2.58)

The collection of charged particles represented by the first two terms form both the material part of the cavity and quantum emitters. We now assume that the cavity–emitter system is well-described within the quasistatic approximation, which applies when all distances in the problem are significantly smaller than the relevant wavelengths. In this limit, the role of the transversal fields is reduced to free-space QED effects such as Lamb shift and radiative decay, which are not significantly modified by the presence of the cavity. We therefore assume here that the transversal fields are negligible, i.e., $\mathbf{A} = \mathbf{B} = \mathbf{E}_\perp \approx 0$, and the Hamiltonian simply becomes

$$\hat{H} = \sum_i \frac{\hat{\mathbf{p}}_i^2}{2m_i} + \sum_{i>j} \frac{q_iq_j}{4\pi\epsilon_0|\mathbf{r}_i - \mathbf{r}_j|},$$

(2.59)

with the sums over i and j still including all particles in the (nano)cavity as well as the quantum emitters. The Coulomb interaction of the second term contains the longitudinal EM fields. We next separate the particles into several groups: one containing the cavity material, and one for each emitter. We assume that the cavity material is "macroscopic" enough that it responds linearly to external fields [39, 40, 44–47], and can thus be well-described by a collection of bosonic modes with frequencies ω_k and annihilation operators a_k (e.g., corresponding to the "instantaneous" plasmon modes in [47]). The Hamiltonian then becomes

$$\hat{H} = \sum_i \frac{\hat{\mathbf{p}}_i^2}{2m_i} + \sum_{i>j} \frac{q_iq_j}{4\pi\epsilon_0|\mathbf{r}_i - \mathbf{r}_j|} + \sum_k \hbar\omega_k\left(\hat{a}_k^\dagger\hat{a}_k + \frac{1}{2}\right) + \sum_k(\hat{a}_k + \hat{a}_k^\dagger)\sum_j q_j\phi_k(\hat{\mathbf{r}}_j),$$

(2.60)

where the first two terms correspond only to the charges associated with the quantum emitters and not with the cavity. The following two terms correspond to the bosonic cavity modes and the interaction of the charges of the emitters (with j running over all the charges of all the emitters) with the electrostatic potential $\phi_k(\mathbf{r})$, i.e., the Coulomb potential corresponding to the charge distribution of each cavity mode. By performing a multipole expansion of the emitter charges, similarly as we did in Sect. 2.2 with a continuous charge distribution, and assuming that the emitter is uncharged and sufficiently localized, this term can be well-approximated by $\hat{\boldsymbol{\mu}} \cdot \hat{\mathbf{E}}(\mathbf{r}_m)$, i.e., the interaction of the emitter dipole with the cavity electric field (the gradient of the potential) at the position $\mathbf{r}_m$ of the molecule, which we write as

$$(\hat{a}_k + \hat{a}_k^{\dagger}) \sum_j q_j \phi_k(\hat{\mathbf{r}}_j) \approx \mathbf{E}_{1\text{ph},k}(\mathbf{r}_m) \cdot \hat{\boldsymbol{\mu}}(\hat{\mathbf{x}}, \hat{\mathbf{R}}), \qquad (2.61)$$

where $\mathbf{E}_{1\text{ph},k}(\mathbf{r}_m) = E_{1\text{ph},k}(\mathbf{r}_m)\epsilon_k$, with polarization vector ϵ, is the single-photon electric field amplitude.

We note that while we have explicitly treated a (nano)cavity within the quasistatic approximation, in which the cavity fields can be understood as due to the instantaneous Coulomb interaction between charged particles, it still makes sense to speak of the cavity modes as electromagnetic or photonic modes with an associated electric field. The modes, which physically correspond to, e.g., plasmonic or phonon-polaritonic resonances, can be seen as strongly confined photons. These modes are most easily obtained by solving Maxwell's equations for a given geometry, either numerically or with approaches such as transformation optics [48]. Only in the limit of extremely small nanocavities does it become possible, and sometimes necessary, to treat them explicitly as a collection of nuclei and electrons using ab initio techniques [49–51].

2.3.2   Common Theoretical Descriptions

While the dipolar Hamiltonian in Eq. (2.21) includes everything required to fully solve the cavity–matter system, this is often too complex to be exactly solved. Some approximations and assumptions are needed to sufficiently simplify the system in order to treat it. Over the years, this has lead to a plethora of theoretical frameworks that we can use in order to describe the physics inside a cavity. In the following we overview some models and approximations that are commonly used in the context of quantum optics and of particular relevance to the content of this thesis.

2.3.2.1   Single-Mode Hamiltonian and Dipole Self-energy

The fundamental Hamiltonian of Eq. (2.21) accounts for the light–matter interaction within the dipole approximation. This couples matter to *all* electromagnetic modes of the system, independently of the properties of the cavity. However, in many hybrid light–matter realizations, the number of relevant electromagnetic modes is reduced to a few or even only one. For instance, in the example of a planar cavity, the distance L between the mirrors only allow photon frequencies with a corresponding wavelength $\lambda_n = 2L/n$, with $n = 1, 2, 3 \ldots$. In many situations it can be safely assumed that the emitters couple to only one photonic frequency to which they are resonant, while the effect of the rest of modes is neglected. This is the so-called *single-mode approximation*, which is a central assumption for many results of this thesis. The importance of this consideration in the field of quantum optics has recently attracted attention, with many studies discussing its validity [52–54].

The assumption of only one photonic mode has a deep impact in the dipolar Hamiltonian of Eq. (2.21), and its consequences must be studied for the particular cavity geometry. Again, in the case of a simple planar cavity, where the electric field is described by Eq. (2.56), the Hamiltonian in the single-mode approximation can be written as (for the detailed deduction see [55])

$$\hat{H} = \hbar\omega_c \hat{a}^\dagger \hat{a} + \sum_i \hat{H}_i + \hat{H}_{dd} + \sqrt{\frac{\hbar\omega_c}{2\epsilon_0 V}} \left(\hat{a}^\dagger + \hat{a} \right) \mathbf{e}_E \cdot \sum_i \hat{\boldsymbol{\mu}}_i + \frac{1}{2\epsilon_0 V} \left(\mathbf{e}_E \cdot \sum_i \hat{\boldsymbol{\mu}}_i \right)^2 ,$$
(2.62)

where $V = LA$ is the effective volume of the relevant mode, and $\mathbf{e}_E$ the unitary vector describing the direction of the electric field. The index i sums over all emitters, which also have dipole–dipole and dipole–induced-dipole interactions represented by $\hat{H}_{dd}$. The last term is the so-called dipole *self-energy* term and represents the self-coupling of the matter to its own field [1, 2, 56]. This coupling is mediated through the omitted high-frequency EM modes, which effectively can influence relevant dipole interactions.

The dipole self-energy term often represents only a small constant energy contribution in second-order perturbation theory, and therefore is usually considered unimportant and is neglected in most common theoretical models in quantum optics. However, in recent years, its proper inclusion to the light–matter Hamiltonian has become a very active topic of discussion [52, 54, 56–61]. For instance, it has been found that when reaching ultra-strong coupling conditions this term is required in order to fit the theory with the experimental data [62]. Nevertheless, for most strong coupling realizations this term can be safely removed, and therefore it is not included in the various theoretical calculations throughout this thesis. Moreover, many strong coupling realizations rely on nanocavities that achieve very strong field concentrations, such as small plasmon- or phonon-polariton nanoantennas and nanoresonators. Particularly, these cavities are the only currently available systems that can obtain few-emitter strong coupling for "real" molecules [63–67]. As discussed above, in these cavities the light–matter interaction is purely longitudinal and, as it is well-known in the literature of macroscopic QED [41], the dipole self-energy term does not appear when the quantum emitters interact with longitudinal modes. However, it should be noted that, in this case, Eq. (2.62) is still accurate if the emitter potential (i.e., $\phi_k(\hat{\mathbf{r}}_j)$ in Eq. (2.61)) is renormalized so that it represents the effect of the high-frequency modes,[12] thus avoiding double counting of modes [55].

2.3.2.2 Tavis–Cummings Model

The single-mode Hamiltonian constitutes the starting point for many theoretical descriptions of cavity QED. Now we wish to model the interaction of N identical emitters characterized as two-level systems with a single cavity mode. For simplicity we assume that there is no direct interaction between the emitters, for example assuming large distances between them. They only collectively interact with the cavity mode. In this treatment we disregard terms that often only contribute as global shifts in energy without affecting the excitation transition mechanism, such as the permanent dipole of the emitters ($\langle g|\hat{\mu}|g\rangle = 0$) or the ground-state energies of both the cavity and the emitters. Finally, from now on, we use atomic units unless stated otherwise ($4\pi\epsilon_0 = \hbar = m_e = e = 1$, with electron mass m_e and elementary charge e). With these considerations we can write the following Hamiltonian:

$$\hat{H} = \omega_c \hat{a}^\dagger \hat{a} + \sum_i^N \left[\omega_e \hat{\sigma}_i^\dagger \hat{\sigma}_i + g_i \left(\hat{\sigma}_i^\dagger + \hat{\sigma}_i \right) \left(\hat{a}^\dagger + \hat{a} \right) \right], \qquad (2.63)$$

where we define the creation and annihilation operators for the two-level system

$$\hat{\sigma}^\dagger = |e\rangle\langle g|, \qquad \hat{\sigma} = |g\rangle\langle e|. \qquad (2.64)$$

Note that the full light–matter coupling is encoded in the parameter $g_i = \mathbf{E}_{1\text{ph}} \cdot \boldsymbol{\mu}_{eg}$, which depends both on the single-photon electric field amplitude $\mathbf{E}_{1\text{ph}}$ and the transition dipole moment from ground to excited state $\boldsymbol{\mu}_{eg}$. Equation (2.63) is known as the *Tavis–Cummings* (TC) or *Dicke* Hamiltonian [68, 69]. For the case of a planar cavity, the electric field amplitude is given by $\mathbf{E}_{1\text{ph}} = \sqrt{\frac{\hbar\omega_c}{2\epsilon_0 V}} \mathbf{e}_E$. It is then straightforward to see that Eq. (2.63) corresponds to the single-mode Hamiltonian where both the dipole–dipole interactions and the dipole self-energy term have been neglected.

The Hamiltonian in Eq. (2.63) contains terms that do not conserve the total number of excitations of the system, namely $\hat{\sigma}_i^\dagger \hat{a}^\dagger$ and $\hat{\sigma}_i \hat{a}$. By transforming the Hamiltonian into the interaction picture, we find that the interaction term now reads

---

[12]This also includes the emitter–emitter interactions in the multiple-emitter case.

$$\hat{H}_{\text{int}} = \sum_i g_i \left[\hat{a}^\dagger \left(\hat{\sigma}_i e^{i(\omega_c - \omega_i)t} + \hat{\sigma}_i^\dagger e^{i(\omega_c + \omega_i)t} \right) + \hat{a} \left(\hat{\sigma}_i e^{i(-\omega_c - \omega_i)t} + \hat{\sigma}_i^\dagger e^{i(-\omega_c + \omega_i)t} \right) \right].$$

$$(2.65)$$

Note that the terms that do not conserve the number of excitations oscillate with frequencies $\omega_c + \omega_i$, much faster than the detuning frequency $\delta = \omega_c - \omega_i$ at which the other terms oscillate. If the couplings g_i are small enough the fast dynamics are not appreciable and quickly average to zero. Neglecting these terms constitutes the so-called *rotating wave approximation* (RWA) [1] and the resulting Hamiltonian now conserves the total number of excitations, i.e. $\left[\hat{H}, \hat{n} \right] = 0$ for $\hat{n} = \hat{a}^\dagger \hat{a} + \sum_i \hat{\sigma}_i^\dagger \hat{\sigma}_i$ and

$$\hat{H} = \omega_c \hat{a}^\dagger \hat{a} + \sum_i^N \left[\omega_e \hat{\sigma}_i^\dagger \hat{\sigma}_i + g_i \left(\hat{\sigma}_i^\dagger \hat{a} + \hat{\sigma}_i \hat{a}^\dagger \right) \right]. \tag{2.66}$$

Since this Hamiltonian conserves the number of excitations, we can analyze the system within the subspace of interest. We can now analyze the collective effects in the single-excitation subspace, which determines the linear properties of the system, e.g. absorption under not too strong driving.[13] It is possible to define the collective operator [70]

$$\hat{S}^\dagger = \frac{1}{\sqrt{\sum_i^N g_i^2}} \sum_i g_i \hat{\sigma}_i^\dagger, \tag{2.67}$$

with which the whole light–matter Hamiltonian can be simply rewritten as

$$\hat{H} = \omega_c \hat{a}^\dagger \hat{a} + \omega_e \hat{S}^\dagger \hat{S} + \frac{\Omega_R}{2} \left(\hat{S}^\dagger \hat{a} + \hat{S} \hat{a}^\dagger \right), \tag{2.68}$$

where we have defined $\Omega_R = 2\sqrt{\sum_i^N g_i^2}$. This quantity is the *Rabi splitting*, a crucial magnitude in strong coupling that will be discussed more in depth in the last part of this section. Note that the collective operators allow us to reduce the system to a 2×2 Hamiltonian, analogous to the scenario of a single-emitter coupled to a single light mode. However in this case, instead of coupling the excited state of one emitter, we have the so-called *bright state* defined as $|B\rangle = \hat{S}^\dagger |0\rangle$, where $|0\rangle$ is the vacuum state. This superposition of excited states collectively couples to the light mode, and the resulting eigenstates of the system (the polaritons) are thus a superposition of bright state and cavity mode. The remaining $N - 1$ states orthogonal to $|B\rangle$ that Eq. (2.66) include but are not described in Eq. (2.68) are known as *dark states*, and are completely uncoupled from the light mode. Even in configurations with many photonic modes (e.g., planar cavities), more than one emitter state is coupled to the photonic mode (typically at low in-plane momentum), but there remain many uncoupled (dark) modes at higher in-plane momentum [71, 72]. These states have energies identical to the uncoupled emitters, $\omega_{DS} = \omega_e$, obscuring further the actual

---

[13]The exact diagonalization of Eq. (2.63) can be done without invoking the RWA. However, for the sake of brevity we present the diagonalization within the single-excitation subspace.

nature of the dark modes, with discussions on whether they are actually affected by strong coupling, or whether they should be thought of as completely unmodified emitter states. In this thesis we shed light on this problem by including the internal rovibrational structure of the molecules in the strong coupling description.

One importance consequence of the TC Hamiltonian is that the Rabi splitting depends on the sum of all the individual emitter–cavity couplings. If we picture the simple scenario where we have N identical emitters in a uniform electric field, we have $g_i = g$ for all emitters, and thus $\Omega_R = 2\sqrt{N}g$. The Rabi splitting is enhanced when an ensemble of emitters collectively interact with the cavity mode. This is one of the key features of *collective strong coupling*. The collection of emitters couple through the bright mode to the cavity with a resulting enhancement of the eigenmode energy splitting of the system of $\sim \sqrt{N}$. Virtually all current experimental strong coupling realizations are achieved by collective strong coupling, using large ensembles of emitters, making the density of emitter dipoles one of the key magnitudes in strong coupling.

2.3.2.3   Extension to More Complex Emitters

The description for the light–matter Hamiltonian discussed above is useful when the emitters can be well-characterized by two-level systems. In the case of organic molecules this is not always true, as they present a plethora of various internal DoF, often resulting in very complex internal structures. Furthermore, this simplified description becomes useless when trying to describe intricate chemical processes. When describing molecules, it is necessary to include a more complete description of the level structure. For instance, the *Holstein–Tavis–Cummings Hamiltonian* [73–79] is a generalization of Eq. (2.63) that treats nuclear motion as harmonic oscillator eigenstates, allowing diagonalization of the full bare-molecule Hamiltonian. This is especially useful for describing phenomena close to the equilibrium position, where the electronic PESs are well-characterized by harmonic potentials, and has been successfully implemented in numerous studies for molecular processes in strong coupling [80]. However, in some phenomena that are far from equilibrium, such as in chemical reactions, the assumption of harmonic electronic PES is not valid, and other methods are necessary.

Another approach consists in extending density functional theory to also include photonic DoF, leading to a quantum-electrodynamical density functional theory [81–84]. This would allow for numerically feasible ab initio simulations of complex correlated light–matter systems, where instead of solving the full matter–photon wavefunction, a set of approximate self-consistent equations of motion for specific quantities can be solved. The main challenge of this powerful idea relies on developing suitable functionals that describe light–matter interactions based on the electron–photon density.

A more complete description can be achieved by using the Born–Oppenheimer approximation in molecules. This allows to describe the system as a collection of independent electronic states characterized by PESs. As discussed in Sect. 2.2, this is standard procedure in organic molecules in chemistry. However, the cavity introduces new degrees of freedom that have to be somehow included in this approximation. This is the main focus of Chaps. 3 and 4, in which we discuss how to perform this adiabatic separation in a light–matter coupled system. We do this by describing photons as discrete DoF and on equal footing as the electrons of the system, thus separating these two coordinates together from the nuclei [85]. This allows a similar description to the one of the Tavis–Cummings Hamiltonian in which we add an explicit dependence on the nuclear configuration to the molecular energy and transition dipole moment. The detailed discussion of this theory and the validity of the BOA in electronic strong coupling are some of the main results of this thesis. The idea is first explored in Chap. 3, where first-principles molecular models are exploited to test the validity of such adiabatic approximation. Then in Chap. 4 we formalize this theory and present a generalization for an arbitrary number of molecules.

We note that another type of adiabatic separation is possible by treating the photons on equal footing as the nuclear DoF. In the Born–Opennheimer approximation this means to treat the photonic mode as a continuous coordinate, and separating it together with the nuclei from the electronic DoF. This is known as the cavity Born–Oppenheimer approximation [82, 86], and because the separation is performed with the low-energy nuclear motion, it works better for ground-state molecules coupled to low-energy photonic modes such as in vibrational strong coupling. We present this approximation and discuss its validity in detail in Chap. 6.

2.3.3 From Weak to Strong Light–Matter Coupling

Finally, let us discuss the different regimes of light–matter interaction depending on the strength of the coupling. The rate of energy exchange between the quantum emitter and the electromagnetic field increases with the coupling, and allows us to differentiate between two distinct regimes of interaction: the weak and strong coupling regimes.

2.3.3.1 Weak Coupling Regime and Purcell Effect

In the *weak coupling regime* the light and matter energy exchange is slower than the decay rate of one of the constituents. Most light–matter interactions in nature occur in this regime, where the electromagnetic field is not confined and thus the coupling strength is very small. This means that the interaction term between quantum emitters and the electromagnetic field can be treated perturbatively and thus approaches like Fermi's Golden rule are applicable (see Sect. 2.2). One important consequence of the presence of the cavity is that it reshapes the density of states of the electromagnetic environment. This can strongly impact the decay rate given by Fermi's Golden rule in Eq. (2.55) by enhancing the spontaneous emission rate of the emitter. This phenomenon is known as the *Purcell effect*, in which the lifetime of the emitter is decreased by adding an additional decay channel. This can improve emission of the quantum emitters achieving, e.g., single-photon sources with impressive figures of merit in the solid state [87].

The modification of the lifetime of the emitter inside the cavity with respect to outside the cavity is controlled by the Purcell factor $\tau_0/\tau \propto F_P$ [88], a key figure of merit in nanophotonics. Different cavity-modified electromagnetic environments lead to different Purcell factors, as F_P depends on the properties of the cavity. In particular, the Purcell factor goes as

$$F_P \propto \frac{Q}{V_{\text{eff}}}, \tag{2.69}$$

where Q is the *quality factor* of the cavity, which quantifies the sharpness of the relevant cavity mode through the ratio between the mode frequency ω_c and its linewidth κ_c, $Q = \omega_c/\kappa_c$, and V_{eff} is the *effective mode volume* corresponding to this same mode. The quality factor describes how good the cavity is in terms of mode lifetime $\tau_c = 1/\kappa_c$, i.e., how long it traps a photon. This allows the emitter to potentially reabsorb a photon it just emitted.

Let us now discuss the light–matter coupling constant g, which is key to characterize the weak coupling regime in theoretical frameworks such as the Tavis–Cummings model (Eq. (2.63)). In general, the coupling constant is directly proportional to the electric field amplitude and the transition dipole moment of the emitter, and can be defined as

$$g_k(\mathbf{r}) = \sqrt{\frac{2\pi\omega_k}{V_{\text{eff}}(\mathbf{r})}} \mathbf{e}_E \cdot \boldsymbol{\mu}_{eg}, \tag{2.70}$$

for the k-th EM mode. For the particular case of a planar mirror microcavity the mode volume will be $V_{\text{eff}} \approx LA$, however in general it can be defined as [89]

$$V_{\text{eff}}(\mathbf{r}) = \frac{\int d^3\mathbf{r}\epsilon(\mathbf{r})|\mathbf{E}(\mathbf{r})|^2}{\epsilon(\mathbf{r})|\mathbf{E}(\mathbf{r})|^2}, \tag{2.71}$$

where $\mathbf{E}(\mathbf{r})$ is the electric field, and therefore V_{eff} effectively depends on the position of the emitter. The mode volume represents how confined is the light in the cavity. In Eq. (2.70) we see that in order to maximize the coupling strength for a particular frequency ω_k we need very large emitter dipoles that align with a very confined electric field. It should be noted that the normalization integral of Eq. (2.71) formally diverges for lossy modes and a more general definition should properly take this into account [47, 90–92].

The light–matter coupling characterizes the energy exchange between light and matter. The larger g, the shorter this exchange process takes. When this is faster that the typical lifetime of either constituent, i.e. if $g \gg \gamma, \kappa_c$, with γ the linewidth of the emitter resonance, we enter the *strong coupling regime*. A photon emitted by the matter constituent is reabsorbed and re-emitted several times before it finally leaks out of the cavity. This can be achieved by improving the cavity through an increase of Q, i.e., enhancing the lifetime of the EM mode inside the cavity, or by decreasing V_{eff}, i.e., confining the electric field and thus increasing its strength.[14] Therefore the Purcell factor becomes a very important quantity to optimize for the achievement of strong coupling. Unfortunately, F_P is very difficult to enhance arbitrarily as typically cavities with smaller mode volumes strongly restrict the quality factor, and vice versa [42].

2.3.3.2 Strong Coupling: A Simple Picture

In order to analyze the strong coupling regime let us present a very simplified model in which we consider a single two-level emitter coupled to a single-mode cavity within the RWA, described by the Jaynes–Cummings model [68], the single-emitter version of Eq. (2.66).[15] Moreover, we will focus on the low-pumping regime so that we can study only the single-excitation subspace, thus reducing the Hilbert space of the problem to only two states: an excited emitter $|e\rangle$ while the cavity is in the vacuum state $|0\rangle$, and a cavity photon $|1\rangle$ while the emitter is in its ground state $|g\rangle$. This leads to the 2×2 Hamiltonian in the basis $\{|e, 0\rangle, |g, 1\rangle\}$ that reads

$$\hat{H} = \begin{pmatrix} \omega_e & g \\ g & \omega_c \end{pmatrix}, \tag{2.72}$$

where $g = \mathbf{E}_{1ph} \cdot \boldsymbol{\mu}_{eg}$ is the coupling constant and $\mathbf{E}_{1ph}$ the single-photon electric field amplitude. As presented above, the losses of the system have a great relevance in the definition of strong coupling and we wish to include them in this simple model. From a theoretical point of view, the losses represent external degrees of freedom not directly represented in our Hamiltonian but that couple to the system and can irreversibly affect it. We can model this loss of energy by adding an imaginary part to the energies of each constituent.[16] We thus set

$$\omega_e = \omega_0 - i\gamma/2; \qquad \omega_c = \omega_0 - i\kappa/2, \tag{2.73}$$

where we consider the system to be in resonance at energy ω_0. If we now diagonalize the system we see that the new eigenvalues are

$$E_\pm = \frac{\omega_e + \omega_c}{2} \pm \frac{1}{2}\sqrt{(\omega_e - \omega_c)^2 + 4|g|^2}. \tag{2.74}$$

---

[14]Note that this also depends on the choice of quantum emitter, as very large Q factors are not useful if the emitter has a broader linewidth than the cavity.

[15]This is equivalent to treating the many-particle TC Hamiltonian with collective operators, thus representing the coupling between the bright state and the cavity mode.

[16]This is equivalent to consider a Lindblad master equation approach and neglecting the excitation refilling terms, making the ground state a population reservoir [93, 94].

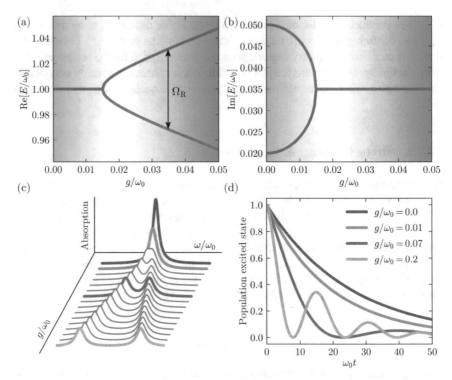

Fig. 2.4 Simple representation of the weak and strong coupling regimes. **a** Real part of the eigenenergies and **b** decay rates of the eigenstates (polaritons) of the Hamiltonian as a function of the coupling strength. **c** View of the energy splitting in the absorption spectrum for different values of the coupling strength. **d** Population dynamics for the stated initiated in the excited emitter state $|e\rangle$. The decay rates are $\gamma/\omega_0 = 0.04$ and $\kappa/\omega0 = 0.1$

Note that due to the complex energies we can achieve negative values inside the square root, so that the eigenenergies $E_\pm$ can experience changes only in the imaginary part, effectively resulting in a modification of the decay rate.

In Fig. 2.4a, b we represent the new energy and lifetime of the resulting eigenstates as a function of the coupling g, for $\gamma/\omega_0 = 0.04$ and $\kappa/\omega0 = 0.1$. These correspond to the real and imaginary part of $E_\pm$. We see that initially (blue region) only the imaginary part is modified while the real part remains unaffected. This is a basic reproduction of the Purcell effect, where the lifetime $\tau \sim 1/\gamma$ of the emitter is decreased in the weak coupling regime. We can see this in the absorption spectrum in Fig. 2.4c, reproduced with a sum of Lorentzians, where for small couplings the width of the peak broadens. In Fig. 2.4d we represent the population dynamics of the system for the initial state $|\Psi(t = 0)\rangle = |e\rangle$. This is obtained by solving the Schrödinger equation $i\partial_t|\Psi(t)\rangle = \hat{H}|\Psi(t)\rangle$ for Eq. (2.72). For $g/\omega_0 = 0$ the emitter does not interact with the cavity and the population decay is governed by the loss rate γ. As the coupling increases the decay rate of the emitter is modified and the population decays faster to the ground state (see green curve for $g/\omega_0 = 0.01$). The highest emission rate out of the cavity is achieved just before entering the strong coupling regime [95].

As coupling increases further, we start seeing a splitting in the real part of the energies, indicating the onset of the strong coupling regime (red region). The new eigenstates of the system are the *upper* (+) and *lower* (−) *polaritons*. While in Fig. 2.4a the two states split in energy for small couplings, this cannot be seen in the absorption spectrum due to the linewidths of each peak. As the coupling is increased, so is the energy separation and the states are properly resolved in absorption (Fig. 2.4c). This energy separation $E_+ - E_- = \Omega_R$ is the so-called *Rabi splitting*, which is the natural frequency of coherent energy exchange between cavity and emitter. This is illustrated in the yellow curve of Fig. 2.4d, where a characteristic oscillatory behavior with frequency Ω_R can be observed for the excited emitter. These are known as *Rabi oscillations*, and they are the signature of cavity and emitter exchanging excitation, indicating that they are no longer the eigenstates of the system [1, 96].

While this an oversimplified description of the phenomenon of strong coupling, it already captures the key physical elements that characterizes this regime. If no losses are considered, Rabi oscillations are always observed. However, these are not reasonable scenarios since in realistic systems the criterion of achieving strong coupling is that the Rabi frequency overcomes the typical decay timescales of the emitter and the cavity. In experiments the usual criterion for the onset of strong coupling is that the absorption peaks are well-resolved. Nevertheless, it should be noted that in more complex systems the crossover between weak and strong coupling regimes may not be so well-defined. Throughout this thesis we do not explicitly include losses in our descriptions of light–matter interaction, so we always theoretically achieve eigenmode splitting. However, we do not ignore the role of the losses, as we discuss their effects in our results.

2.4 Summary of Methods Applied in This Thesis

The goal of this final section is to provide a brief summary of some of the methods and theoretical techniques used throughout this thesis. The fundamental pillar of most results is the many-molecule Hamiltonian coupled to a light mode:

$$\hat{H} = \omega_c \hat{a}^\dagger \hat{a} + \sum_i^N \left(\hat{H}_{\text{mol}}^{(i)} + \mathbf{E}_{1\text{ph,i}} \cdot \hat{\boldsymbol{\mu}}_i (\hat{a}^\dagger + \hat{a}) \right), \tag{2.75}$$

where the the dipole–dipole interaction between molecules is usually disregarded, with the exception of Sect. 6.5. Additionally, we do not include the dipole self-energy term. For most calculations this can be safely removed, as its contribution is of higher order than the linear dipolar interaction considered. Moreover, this term does not appear in the interaction with cavities that can be described within the quasistatic approximation, such as plasmonic cavities, so widely used to achieve few-molecule strong coupling.

In the following two chapters we explore the foundations of a theory of polaritonic chemistry. In particular, in Chap. 3 we analyze the validity of the Born–Oppenheimer approximation, for which the absorption spectrum of the system in and out of strong coupling is calculated. This is done by solving the corresponding system Hamiltonian and then introducing the resulting eigenstates in the scattering amplitude of Eq. (2.54). Then, in Chap. 4, we directly tackle how to solve the Hamiltonian in Eq. (2.75) for large number of molecules. This is achieved using collective spin operators, such as the ones used in the TC model Eq. (2.67).

In Chap. 5 we solve again the Hamiltonian in Eq. (2.75) for specific molecular models that present chemical reactions. In particular, this Hamiltonian is used to calculate the time evolution of a wavepacket on a PES using the time-independent Schrödinger equation:

$$i\partial_t |\psi(t)\rangle = \hat{H}_{\text{tot}} |\psi(t)\rangle. \tag{2.76}$$

Furthermore, methods often used in chemistry (see subsection "Chemical processes" in Sect. 2.2) are exploited in this chapter. For example, in order to explore reaction pathways, the MEP of the system is calculated using the nudged elastic band method [97]. Additionally, TST is used to calculate reaction rates in the various processes analyzed.

Then, in Chap. 6 we again use Eq. (2.75) for the case of nanoscale cavities. In order to analyze the reaction rates of a molecular model, we compute the exact quantum reaction rate of Eq. (2.42), which is formally equivalent to solving Eq. (2.76) for an averaged thermal distribution as initial state. These rates are also compared to the ones computed using TST. In the theory we develop we apply the cavity Born–Oppenheimer approximation [82, 86], and present an extension of the discussion of the subsection "Born–Oppenheimer approximation" of Sect. 2.2. Furthermore, in order to calculate the ground state of the system we apply perturbation theory to the Hamiltonian Eq. (2.75), and we discuss the importance of the static polarizability of Eq. (2.52).

Finally, we note that in all the following chapters, atomic units (a.u.) are often used $(4\pi\epsilon_0 = \hbar = m_e = e = 1$, with electron mass m_e and elementary charge e). In particular, spatial DoF such as nuclear coordinates are expressed in atomic units of length, which are defined as the Bohr radius, i.e., 1 a.u. ≈ 0.529 Å.

References

1. Cohen-Tannoudji C, Dupont-Roc J, Grynberg G (1997) Photons and atoms
2. Steck DA (2007) Quantum and atom optics, vol 47
3. Dirac PAM (1927) The quantum theory of the emission and absorption of radiation. Proc R Soc A Math Phys Eng Sci 114:243
4. Dirac PAM (1981) The principles of quantum mechanics, vol 27. Oxford University Press

5. Majety VP, Zielinski A, Scrinzi A (2014) Mixed gauge in strong laser-matter interaction. J Phys B At Mol Opt Phys 48:025601
6. Göppert-Mayer M (1931) Über Elementarakte mit zwei Quantensprüngen. Annalen der Physik 401:273
7. Power EA, Zienau S (1959) Coulomb gauge in non-relativistic quantum electro-dynamics and the shape of spectral lines. Philos Trans R Soc Lond Ser A Math Phys Sci 251:427
8. Woolley RG (1971) Molecular quantum electrodynamics. Proc R Soc Lond A 321:557
9. Power E, Thirunamachandran T (1982) Quantum electrodynamics in a cavity. Phys Rev A 25:2473
10. Power EA, Thirunamachandran T (1980) The multipolar Hamiltonian in radiation theory. Proc R Soc Lond A Math Phys Sci 372:265
11. Born M, Oppenheimer R (1927) Zur Quantentheorie der Molekeln. Annalen der Physik 389:457
12. Tully JC (2000) Perspective on "Zur Quantentheorie der Molekeln". Theor Chem Acc Theory Comput Model (Theoretica Chimica Acta) 103:173
13. Born M, Huang K (1954) Dynamical theory of crystal lattices. Clarendon Press
14. Worth GA, Cederbaum LS (2004) Beyond Born-Oppenheimer: molecular dynamics through a conical intersection. Annu Rev Phys Chem 55:127
15. Levine BG, Martínez TJ (2007) Isomerization through conical intersections. Annu Rev Phys Chem 58:613
16. Mead CA, Truhlar DG (1982) Conditions for the definition of a strictly diabatic electronic basis for molecular systems. J Chem Phys 77:6090
17. Novotny L, Hecht B (2009) Principles of nano-optics. ISBN: 9781107005464
18. Leonas V et al (2000) Photosynthetic excitons. World Scientific
19. Weiss U (2012) Quantum dissipative systems, vol 10. World Scientific
20. Stone A (2013) The theory of intermolecular forces, 2nd edn. OUP Oxford
21. Sheppard D, Terrell R, Henkelman G (2008) Optimization methods for finding minimum energy paths. J Chem Phys 128:134106
22. Truhlar DG, Gordon MS (1990) From force fields to dynamics: classical and quantal paths. Science 249:491
23. Prezhdo OV, Rossky PJ (1997) Mean-field molecular dynamics with surface hopping. J Chem Phys 107:825
24. Barbatti M (2011) Nonadiabatic dynamics with trajectory surface hopping method. Wiley Interdisc Rev Comput Mol Sci 1:620
25. Eyring H (1935) The activated complex in chemical reactions. J Chem Phys 3:107
26. Arrhenius S (1889) Über die dissociationswärme und den einfluss der temperatur auf den dissociationsgrad der elektrolyte. Zeitschrift für physikalische Chemie 4:96
27. Miller WH (1993) Beyond transition-state theory: a rigorous quantum theory of chemical reaction rates. Acc Chem Res 26:174
28. Yamamoto T (1960) Quantum statistical mechanical theory of the rate of exchange chemical reactions in the gas phase. J Chem Phys 33:281
29. Miller WH (1974) Quantum mechanical transition state theory and a new semiclassical model for reaction rate constants. J Chem Phys 61:1823
30. Miller WH, Schwartz SD, Tromp JW (1983) Quantum mechanical rate constants for bimolecular reactions. J Chem Phys 79:4889
31. Jackson JD (2007) Classical electrodynamics. Wiley
32. Bonin KD, Kresin VV (1997) Electric-dipole polarizabilities of atoms, molecules, and clusters. World Scientific Publishing Co. Pte. Ltd
33. Rescigno TN, McKoy V (1975) Rigorous method for computing photoabsorption cross sections from a basis-set expansion. Phys Rev A 12:522
34. Kasha M (1950) Characterization of electronic transitions in complex molecules. Discuss Faraday Soc 9:14
35. Nitzan A, Jortner J, Rentzepis PM (1971) Internal conversion in large molecules. Mol Phys 22:585

36. May V, Kühn O (2011) Charge and energy transfer dynamics in molecular systems. Wiley-VCH Verlag GmbH & Co. KGaA, Weinheim, Germany
37. Nitzan A (2006) Chemical dynamics in condensed phases: relaxation, transfer and reactions in condensed molecular systems. Oxford University Press
38. Barbry M, Koval P, Marchesin F, Esteban R, Borisov AG, Aizpurua J, Sánchez-Portal D (2015) Atomistic near-field nanoplasmonics: reaching atomic-scale resolution in nanooptics. Nano Lett 15:3410
39. Huttner B, Barnett SM (1992) Quantization of the electromagnetic field in dielectrics. Phys Rev A 46:4306
40. Scheel S, Buhmann SY (2008) Macroscopic quantum electrodynamics—concepts and applications. Acta Physica Slovaca 58:675
41. Buhmann SY (2007) Casimir-polder forces on atoms in the presence of magnetoelectric bodies. Thesis (PhD), Friedrich-Schiller-Universität Jena
42. Kavokin A, Baumberg J, Malpuech G, Laussy F (2008) Microcavities
43. Tame MS, McEnery K, Özdemir Ş, Lee J, Maier S, Kim M (2013) Quantum plasmonics. Nat Phys 9:329
44. Hopfield J (1958) Theory of the contribution of excitons to the complex dielectric constant of crystals. Phys Rev 112:1555
45. Dung HT, Knöll L, Welsch D-G (1998) Three-dimensional quantization of the electromagnetic field in dispersive and absorbing inhomogeneous dielectrics. Phys Rev A 57:3931
46. Van Vlack C, Kristensen PT, Hughes S (2012) Spontaneous emission spectra and quantum light-matter interactions from a strongly coupled quantum dot metal-nanoparticle system. Phys Rev B 85:075303
47. Alpeggiani F, Andreani LC (2014) Quantum theory of surface plasmon polaritons: planar and spherical geometries. Plasmonics 9:965
48. Li R-Q, Hernángomez-Pérez D, García-Vidal FJ, Fernández-Domínguez AI (2016) Transformation optics approach to plasmon-exciton strong coupling in nanocavities. Phys Rev Lett 117:107401
49. Savage KJ, Hawkeye MM, Esteban R, Borisov AG, Aizpurua J, Baumberg JJ (2012) Revealing the quantum regime in tunnelling plasmonics. Nature 491:574
50. Zhang P, Feist J, Rubio A, García-González P, García-Vidal FJ (2014) Ab initio nanoplasmonics: the impact of atomic structure. Phys Rev B 90:161407(R)
51. Varas A, García-González P, Feist J, García-Vidal FJ, Rubio A (2016) Quantum plasmonics: from jellium models to ab initio calculations. Nanophotonics 5:409
52. Vukics A, Grießer T, Domokos P (2014) Elimination of the A-square problem from cavity QED. Phys Rev Lett 112:073601
53. Keeling J (2007) Coulomb interactions, gauge invariance, and phase transitions of the Dicke model. J Phys Condens Matter 19:295213
54. Kirton P, Roses MM, Keeling J, Dalla Torre EG (2018) Introduction to the Dicke model: from equilibrium to nonequilibrium, and vice versa. Adv Quantum Technol 1800043
55. De Bernardis D, Jaako T, Rabl P (2018) Cavity quantum electrodynamics in the nonperturbative regime. Phys Rev A 97:043820
56. Rokaj V, Welakuh DM, Ruggenthaler M, Rubio A (2018) Light-matter interaction in the long-wavelength limit: no ground-state without dipole self-energy. J Phys B 51:034005
57. Todorov Y, Sirtori C (2014) Few-electron ultrastrong light-matter coupling in a quantum LC circuit. Phys Rev X 4:041031
58. De Bernardis D, Pilar P, Jaako T, De Liberato S, Rabl P (2018) Breakdown of gauge invariance in ultrastrong-coupling cavity QED. Phys Rev A 98:053819
59. Andrews DL, Jones GA, Salam A, Woolley RG (2018) Perspective: quantum Hamiltonians for optical interactions. J Chem Phys 148:040901
60. Stokes A, Nazir A (2019) Gauge ambiguities imply Jaynes-Cummings physics remains valid in ultrastrong coupling QED. Nat Commun 10:499
61. Kónya G, Vukics A, Domokos P. The equivalence of the Power-Zineau-Woolley picture and the Poincaré gauge from the very first principles

62. George J, Chervy T, Shalabney A, Devaux E, Hiura H, Genet C, Ebbesen TW (2016) Multiple Rabi splittings under ultrastrong vibrational coupling. Phys Rev Lett 117:153601
63. Chikkaraddy R, de Nijs B, Benz F, Barrow SJ, Scherman OA, Rosta E, Demetriadou A, Fox P, Hess O, Baumberg JJ (2016) Single-molecule strong coupling at room temperature in plasmonic nanocavities. Nature 535:127
64. Chikkaraddy R, Turek VA, Kongsuwan N, Benz F, Carnegie C, van de Goor T, de Nijs B, Demetriadou A, Hess O, Keyser UF, Baumberg JJ (2018) Mapping nanoscale hotspots with single-molecule emitters assembled into plasmonic nanocavities using DNA origami. Nano Lett 18:405
65. Autore M, Li P, Dolado I, Alfaro-Mozaz FJ, Esteban R, Atxabal A, Casanova F, Hueso LE, Alonso-González P, Aizpurua J, Nikitin AY, Vélez S, Hillenbrand R (2018) Boron nitride nanoresonators for phonon-enhanced molecular vibrational spectroscopy at the strong coupling limit. Light Sci Appl 7:17172
66. Zengin G, Wersäll M, Nilsson S, Antosiewicz TJ, Käll M, Shegai T (2015) Realizing strong light-matter interactions between single-nanoparticle plasmons and molecular excitons at ambient conditions. Phys Rev Lett 114:157401
67. Gubbin CR, Maier SA, De Liberato S (2017) Theoretical investigation of phonon polaritons in SiC micropillar resonators. Phys Rev B 95:035313
68. Jaynes ET, Cummings FW (1963) Comparison of quantum and semiclassical radiation theories with application to the beam maser. Proc IEEE 51:89
69. Tavis M, Cummings FW (1968) Exact solution for an N-molecule-radiation-field Hamiltonian. Phys Rev 170:379
70. Garraway BM (2011) The Dicke model in quantum optics: Dicke model revisited. Philos Trans R Soc A Math Phys Eng Sci 369:1137
71. Michetti P, Mazza L, La Rocca GC (2015) Strongly coupled organic microcavities. In: Zhao YS (ed) Organic nanophotonics, nano-optics and nanophotonics, vol 39. Springer, Berlin, Heidelberg
72. Agranovich V, Gartstein YN, Litinskaya M (2011) Hybrid resonant organic-inorganic nanostructures for optoelectronic applications. Chem Rev 111:5179
73. Spano FC (2015) Optical microcavities enhance the exciton coherence length and eliminate vibronic coupling in J-aggregates. J Chem Phys 142
74. Herrera F, Spano FC (2016) Cavity-controlled chemistry in molecular ensembles. Phys Rev Lett 116:238301
75. Ćwik JA, Kirton P, De Liberato S, Keeling J (2016) Excitonic spectral features in strongly coupled organic polaritons. Phys Rev A 93:33840
76. Zeb MA, Kirton PG, Keeling J (2018) Exact states and spectra of vibrationally dressed polaritons. ACS Photonics 5:249
77. Wu N, Feist J, Garcia-Vidal FJ (2016) When polarons meet polaritons: exciton-vibration interactions in organic molecules strongly coupled to confined light fields. Phys Rev B 94:195409
78. Herrera F, Spano FC (2017) Dark vibronic polaritons and the spectroscopy of organic microcavities. Phys Rev Lett 118
79. Herrera F, Spano FC (2018) Theory of nanoscale organic cavities: the essential role of vibration-photon dressed states. ACS Photonics 5:65
80. del Pino J (2018) Vibrational and electronic strong light–matter coupling with molecular excitations. Thesis (PhD), Universidad Autónoma de Madrid
81. Flick J, Ruggenthaler M, Appel H, Rubio A (2015) Kohn-Sham approach to quantum electrodynamical density-functional theory: exact time-dependent effective potentials in real space. Proc Natl Acad Sci 112:15285
82. Flick J, Ruggenthaler M, Appel H, Rubio A (2017) Atoms and molecules in cavities, from weak to strong coupling in quantum-electrodynamics (QED) chemistry. Proc Natl Acad Sci 114:3026
83. Tokatly IV (2013) Time-dependent density functional theory for many-electron systems interacting with cavity photons. Phys Rev Lett 110:233001

84. Ruggenthaler M, Flick J, Pellegrini C, Appel H, Tokatly IV, Rubio A (2014) Quantum-electrodynamical density-functional theory: bridging quantum optics and electronic-structure theory. Phys Rev A 90:012508
85. Feist J, Galego J, Garcia-Vidal FJ (2018) Polaritonic chemistry with organic molecules. ACS Photonics 5:205
86. Flick J, Appel H, Ruggenthaler M, Rubio A (2017) Cavity Born-Oppenheimer approximation for correlated electron-nuclear-photon systems. J Chem Theory Comput 13:1616
87. Somaschi N, Giesz V, De Santis L, Loredo J, Almeida MP, Hornecker G, Portalupi SL, Grange T, Antón C, Demory J et al (2016) Near-optimal single-photon sources in the solid state. Nat Photonics 10:340
88. Purcell EM (1946) Spontaneous emission probabilities at radio frquencies. Phys Rev 69:674
89. Louisell WH, Louisell WH (1973) Quantum statistical properties of radiation, vol 7. Wiley, New York
90. Koenderink AF (2010) On the use of purcell factors for plasmon antennas. Opt Lett 35:4208
91. Kristensen PT, Van Vlack C, Hughes S (2012) Generalized effective mode volume for leaky optical cavities. Opt Lett 37:1649
92. Sauvan C, Hugonin JP, Maksymov IS, Lalanne P (2013) Theory of the spontaneous optical emission of nanosize photonic and plasmon resonators. Phys Rev Lett 110:237401
93. Brecha R, Rice P, Xiao M (1999) N two-level atoms in a driven optical cavity: quantum dynamics of forward photon scattering for weak incident fields. Phys Rev A 59:2392
94. Sáez-Blázquez R, Feist J, Fernández-Domínguez A, García-Vidal F (2017) Enhancing photon correlations through plasmonic strong coupling. Optica 4:1363
95. Bozhevolnyi SI, Khurgin JB (2016) Fundamental limitations in spontaneous emission rate of single-photon sources. Optica 3:1418
96. Törmä P, Barnes WL (2015) Strong coupling between surface plasmon polaritons and emitters: a review. Rep Prog Phys 78:13901
97. Henkelman G, Uberuaga BP, Jónsson H (2000) A climbing image nudged elastic band method for finding saddle points and minimum energy paths. J Chem Phys 113:9901

Chapter 3
Molecular Structure in Electronic Strong Coupling

3.1 Introduction

As introduced in Sect. 1.3, organic molecules were initially used in order to achieve robust room-temperature strong coupling, and were merely seen as another method to manipulate light. However, their complex internal structure soon became apparent in many experiments where nuclear degrees of freedom played a relevant role. The necessity of a theory of strong coupling that included the rovibrational structure was undeniable. Such a theory implies having a nucleus–electron–photon coupled system, in which three different timescales play a role. In Sect. 2.3 we have reviewed different attempts to treat such systems, but that lacked the potential to describe complex processes such as chemical reactions. This can be achieved by exploiting the usual picture of potential energy surfaces (PES), so widely-used in chemistry. However, this approach faces the challenge of separating not only electronic and nuclear DoF, but also the photonic one. In here we study in detail the validity of the Born–Oppenheimer approximation for molecule–cavity systems. In order to do that, we introduce a simple first-principles model that fully describes nuclear, electronic, and photonic degrees of freedom, but can be solved without approximations. This allows us to provide a simple picture for understanding the induced modifications of molecular structure. This enables us to analyze the validity of standard approximations in chemistry for our light–matter Hamiltonian.

In Sect. 3.2, after introducing the model, we discuss under which conditions and in which form the Born–Oppenheimer approximation (BOA) [1, 2] is valid in the strong coupling regime for a single molecule. The BOA is widely used in molecular and solid state physics and quantum chemistry, and provides a simple picture of nuclei moving on effective potential energy surfaces generated by the electrons, which underlies most of the current understanding of chemical reactions [2]. However, the BOA depends on the separation of electronic and nuclear energy scales, i.e., the fact that electrons typically move much faster than nuclei. It could thus conceivably break down when an additional, intermediate timescale is introduced under strong coupling to an EM mode. The speed of energy exchange between field and molecules

J. Galego Pascual, *Polaritonic Chemistry*, Springer Theses, https://doi.org/10.1007/978-3-030-48698-3_3

is determined by the Rabi frequency Ω_R, and typical experimental values of hundreds of meV land squarely between typical nuclear ($\simeq 100$ meV) and electronic ($\simeq 2$ eV) energies. We show that the BOA indeed breaks down at intermediate Rabi splittings, but remains valid when Ω_R becomes large enough. For cases where it breaks down, we show that the nonadiabatic coupling terms can be obtained to a good approximation without requiring knowledge of the electronic wave functions.

In Sect. 3.3, we focus on the effects of strong coupling when more than one molecule is involved, using two molecules as the simplest test case. In experiments, strong coupling is achieved by collective coupling to a large number of molecules, under which the Rabi frequency is enhanced by a factor of $\sqrt{N}$. Despite the fact that strong coupling at the single-molecule level has already been achieved experimentally [3], in most cavities the number of molecules is usually much larger, from a few hundred in nanoparticles hosting LSPs [4], to $\gtrsim 10^5$ within planar microcavities [5–8]. In this context, it is well known that only a small fraction of the collective electronic excitations are strongly coupled [9–11], with a large number of "dark" or "uncoupled" modes that show no mixing with the EM mode and no energy shift. We show that even these dark modes are affected by strong coupling, with the nuclear motion of separated molecules becoming correlated. This has later been proven to be of great relevance in exciton transport [12, 13].

3.2   Single Molecule

In this section, we introduce our model for a single molecule coupled to an EM mode. Due to the exponential scaling of the Hilbert space with the number of DoF, solving the full time-independent Schrödinger equation for an organic molecule without the BOA is an extremely challenging task that even modern supercomputers can only handle for very small molecules. We thus employ a reduced-dimensionality model that we can easily solve, both for the bare molecule and after coupling to an EM mode.

3.2.1   Bare Molecule Model

We work within the single-active-electron approximation, in which all but one electron are frozen around the nuclei, and additionally restrict the motion of the active electron to one dimension, x. Furthermore, we only treat one nuclear DoF, the reaction coordinate R. This could correspond to the movement of a single bond in a molecule, but can equally well represent collective motion, e.g., the breathing mode of a carbon ring. The effective molecular Hamiltonian then highly resembles that of a one-dimensional diatomic molecule,

$$\hat{H}_{\mathrm{mol}} = \hat{T}_{\mathrm{n}} + \hat{T}_{\mathrm{e}} + V_{\mathrm{en}}(x; R) + V_{\mathrm{nn}}(R), \tag{3.1}$$

where $\hat{T}_n = \frac{\hat{P}^2}{2M}$ and $\hat{T}_e = \frac{\hat{p}^2}{2}$ are the nuclear and electronic kinetic energy operators (with $\hat{P}$, $\hat{p}$ the corresponding momenta), and M is the nuclear mass. The potentials $V_{en}(x, R)$ and $V_{nn}(R)$ represent the effective electron–nuclei and internuclear interactions, where we assume two nuclei located at $x = \pm R/2$. These potentials encode the information about the frozen electrons as well as the nuclear structure of the molecule, and can be adjusted to approximately represent different molecules.

The electron–nucleus interaction V_{en} contains the interaction of the active electron with each nucleus, as well as with the frozen electrons surrounding it. Assuming a nuclear charge of Z, we have $2Z - 1$ frozen electrons distributed across the two nuclei. For large distances, the active electron should thus feel a Coulomb potential with an effective charge of $\frac{1}{2}$ from each nucleus. Conversely, at very small distances, the active electron is not affected by the cloud of frozen electrons and feels an effective charge of Z. Since we are working within one dimension, we use a soft Coulomb potential to take into account that the electron avoids the singularity at the nucleus. We choose a simple model potential fulfilling these conditions:

$$V_{en}(r) = -\frac{\frac{1}{2} + (Z - \frac{1}{2})e^{-\frac{r}{r_0}}}{\sqrt{r^2 + \alpha^2}}, \qquad (3.2)$$

where α is the softening parameter, r_0 describes the localization of the frozen electrons around the nucleus, and r is the electron-nucleus distance. The total potential is thus $V_{en}(x, R) = V_{en}(|x - R/2|) + V_{en}(|x + R/2|)$, shown in Fig. 3.1a.

The internuclear potential $V_{nn}(R)$ represents the interaction between the nuclei and the $2Z - 1$ frozen electrons, i.e., the ground state potential energy surface of the molecular ion. We model this surface by a Morse potential (see Fig. 3.1b)

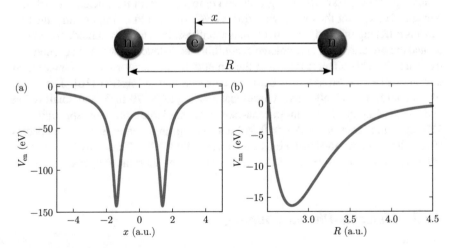

Fig. 3.1 Top: schematic representation of our molecular model. Bottom: model potentials for **a** electron-nuclei and **b** nucleus-nucleus interactions

Table 3.1 Molecular model parameters

	α	Z	r_0	D_e	A	R_0	M
R6G	0.2	1	3.2	0.6	2.5	2.8	10^6
Anthracene	0.271	1	2.2	17.1	0.24	2.88	4.5×10^4

$$V_{nn}(R) = D_e \left(1 - e^{A(R-R_0)}\right)^2, \tag{3.3}$$

which adds three new parameters: the dissociation energy D_e, the equilibrium distance R_0, and the width of the potential well A. By tuning the seven free parameters we have at our disposal ($M, Z, \alpha, r_0, D_e, R_0$ and A), we can approximately fit both the electronic and vibrational structure and absorption spectrum to those of real organic molecules.

We now solve the stationary Schrödinger equation $\hat{H}_{mol}\Psi(x, R) = E\Psi(x, R)$ for the bare-molecule Hamiltonian Eq. (3.1) without further approximations by representing $\hat{H}_{mol}$ on a two-dimensional grid in x and R. We also calculate the independent PES within the BOA and the corresponding nuclear eigenstates. For a bare molecule, the results of solving the Schrödinger equation without approximations and the ones corresponding to the BOA are virtually identical and thus not shown here.

In the following, we will focus on two model molecules, which approximately reproduce the absorption spectra of rhodamine 6G (R6G) and anthracene molecules that are commonly used in experimental realizations of strong coupling [7, 14, 15]. The molecular parameters used are shown in Table 3.1, all expressed in the appropriate atomic units. Only the first two PES, corresponding to the ground $V_g(R)$ and first electronically excited $V_e(R)$ states, play a role in the results discussed in the following. They are shown in Fig. 3.2a and c, together with the nuclear probability densities $|\chi(R)|^2$ for the lowest vibrational levels on each PES. Importantly, the two models differ significantly in two relevant quantities: the vibrational mode frequency ω_v and the offset ΔR, i.e., the change in equilibrium distance between the ground and excited PES. This offset is related to the strength of the electron-phonon interaction and influences the Stokes shift between emission and absorption [16]. The R6G-like model has relatively small vibrational spacing $\omega_v \approx 70$ meV and small offset $\Delta R \approx 0.018$ a.u., while the anthracene-like model has large vibrational spacing $\omega_v \approx 180$ meV and large offset $\Delta R \approx 0.092$ a.u.. Accordingly, their absorption spectra (Fig. 3.2b and d, obtained using Eq. (2.53)) are qualitatively different, with anthracene showing a broader absorption peak with well-resolved vibronic subpeaks.

3.2.2 Molecule-Photon Coupling

We now add a single photonic mode and its coupling to the molecule (within the dipole approximation) into the molecular Hamiltonian. This is achieved through the

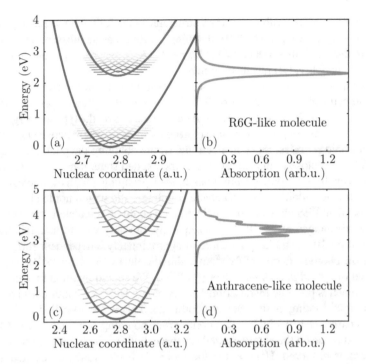

Fig. 3.2 Bare-molecule potential energy surfaces of the two first electronic states in the BOA for **a** the rhodamine 6G-like model molecule and **c** the anthracene-like model molecule. The vibrational levels and associated nuclear probability densities are represented on top of the PES. **b** and **d**: Absorption spectrum for the **b** R6G-like and **d** anthracene-like molecule in arbitrary units

single-mode approximation, which leads to the general Hamiltonian of Eq. (2.62). In the present calculation we focus on the linear dipolar coupling that dominates the light–matter interaction, and thus neglect small energy contributions such as the dipole self-energy, as discussed in detail in Sect. 2.3. The resulting Hamiltonian is

$$\hat{H}_{c-mol} = \hat{H}_{mol} + \omega_c \hat{a}^\dagger \hat{a} + E_{1ph}\hat{\mu}(\hat{a}^\dagger + \hat{a}), \qquad (3.4)$$

where $\hat{\mu}$ is the dipole operator of the molecule ($\hat{\mu} = \hat{x}$ in our case[1]), $\hat{a}^\dagger$ and $\hat{a}$ are the creation and annihilation operators for the bosonic EM field mode, ω_c is its frequency, and E_{1ph} is the coupling strength constant, given by the electric field amplitude (along the x-axis) of a single photon. Note that the Hamiltonian of Eq. (3.4) is analogous to the Jaynes–Cummings Hamiltonian (or alternatively, the single-emitter TC Hamil-

---

[1]Due to nuclear symmetry, the only contribution to the total dipole moment operator is electronic. Including an asymmetry in our model makes the nuclear contribution to the dipole non-zero and thus $\hat{\mu} = \hat{x} + \hat{R}$. This would add a non-zero permanent dipole to the molecule, which could introduce small energy contributions to the PES. We note that this does not change the qualitative analysis of this chapter, and that we include a discussion of its effects on the ground state in Sect. 6.6.

tonian) for an emitter with an arbitrary structure. In the following, we always set the photon energy ω_c to achieve "zero detuning", with ω_c at the absorption maximum of the molecule. This gives $\omega_c \approx V_e(R_{eq}) - V_g(R_{eq})$, where R_{eq} is the equilibrium position at which $V_g(R)$ has its minimum.

Compared to the bare-molecule case, the Hamiltonian now includes a new degree of freedom, the photon number $n \in \{0, 1, 2, \ldots\}$, with the system eigenstates defined by $\hat{H}_{c-mol}\Psi(x, n, R) = E\Psi(x, n, R)$. As discussed above, the typical energies associated with strong coupling in organic molecules are somewhere between the nuclear and electronic energies. Since in electronic strong coupling the photonic frequency is by definition close to the energy of the first excited state, it makes more sense to group it with the electronic Hamiltonian. Indeed, grouping it with the nuclear motion would introduce additional nonadiabatic couplings, and would not lead to a picture of independent PES on which the nuclei motion could be calculated, and would thus ruin the advantages of the BOA. Consequently, the only way to maintain the usefulness of the BOA and keep a picture of approximately independent surfaces is to include the photonic degree of freedom within the electronic Hamiltonian, leading to a new set of hybrid *polaritonic* PESs (PoPESs). We should mention that an alternative approach is possible by describing the photonic DoF as a continuous parameter and on equal footing as the nuclear coordinate. This is the so-called cavity BOA, which is particularly useful for vibrational strong coupling, as we explore in Chap. 6.

We first focus on the singly excited subspace, within which the splitting between polaritons is observed. Here, either the molecule is electronically excited and no photons are present, or the molecule is in its electronic ground state and the photon mode is singly occupied. At zero coupling ($E_{1ph} = 0$), this gives two uncoupled PES ($V_e(R)$ and $V_g(R) + \omega_c$, dashed curves in Fig. 3.3) that cross close to R_{eq} for our choice of ω_c. When the electron–photon coupling is non-zero but small, a narrow

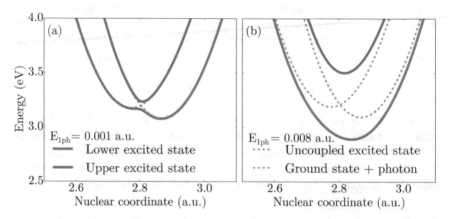

Fig. 3.3 Strongly coupled electronic PES (solid lines) in the singly excited subspace, for the anthracene-like molecule for **a** $E_{1ph} = 0.001$ a.u. and **b** $E_{1ph} = 0.008$ a.u.. The dashed lines show the corresponding uncoupled states: A molecule in the first excited state, $V_e(R)$, and a molecule in the ground state with one photon present, $V_g(R) + \omega_c$

avoided crossing develops instead (solid lines in Fig. 3.3a), while for large coupling strengths, the energy exchange between photonic and electronic degrees of freedom is so fast that we observe two entirely new PES (Fig. 3.3b), the hybrid PoPESs that contain a mixture of electronic and photonic excitation, the hallmark of the strong coupling regime.

As discussed above, the BOA is known to be valid when two PES are sufficiently separated from each other. This implies that the BOA breaks down when E_{1ph} is small and the two PES possess a narrow avoided crossing. This in itself is not a surprising result—when the electron–photon coupling is very small, the system is not even in the strong coupling regime, and the photon mode is better treated as a small perturbation. Fortunately, the weak coupling regime is also not interesting from the standpoint of understanding or modifying molecular structure through strong coupling. The real question thus must be: *How strong* does the electron–photon coupling have to be for the BOA to be valid, and is this condition fulfilled for realistic experimental parameters? In order to better quantify the agreement between the BOA and the full solution, we next turn to an easily measured physical observable: the absorption spectrum.

3.2.3 Absorption

In order to calculate the absorption spectrum that would be observed under driving by an external field, the details of the experimental setup would have to be taken into account. For example, for a planar microcavity, an input-output formalism [17] in which the cavity mode is driven by external photons through the cavity mirrors, would be most appropriate. On the other hand, if the molecules are placed next to a metallic nanoparticle, an external field would typically drive both the molecules and the localized surface plasmon resonance. In the following, we calculate the absorption spectra under the assumption that only the molecules are directly coupled to the external light source. This allows to focus on the influence of the molecular structure on the absorption spectrum, without contamination from a peak due to the essentially pure EM mode at low coupling. We have explicitly checked that our conclusions do not depend on the choice of driving operator. Under these assumptions, the absorption cross section at frequency ω can be calculated using Eq. (2.53) as described in Sect. 2.2, i.e.,

$$\sigma(\omega) = \frac{4\pi\omega}{c}\mathrm{Im}\left[\lim_{\epsilon \to 0}\sum_k \frac{|\langle\Psi_k|\hat{\mu}|\Psi_0\rangle|^2}{\omega_k - \omega_0 - \omega - i\epsilon}\right], \tag{3.5}$$

where the sum runs over all eigenstates $|\Psi_k\rangle$ of the system with energies ω_k, and with $|\Psi_0\rangle$ the overall ground state. As we do not include incoherent processes in our calculation, this would give δ-like peaks in the absorption cross section. In the plots shown in the following, we instead introduce a phenomenological width representing losses and pure dephasing by setting ε to a small non-zero value, such that the

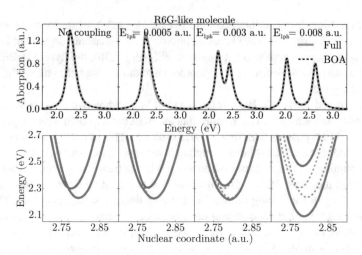

Fig. 3.4 Top row: Absorption cross sections of a single R6G-like model molecule, calculated using the full Hamiltonian without approximation (solid green lines) and within the BOA (dashed black lines) for several values of the coupling strength. Bottom row: Corresponding single-molecule PES in the single-excitation subspace in strong coupling (solid lines) and uncoupled (dashed lines)

absorption cross section becomes a sum of Lorentzians. For the bare-molecule case without coupling to an EM mode, the absorption spectra of our two model molecules approximately agree with those of R6G (Fig. 3.2b, [15]) and anthracene (Fig. 3.2d, [7]).

In the upper rows of Figs. 3.4 and 3.5 we compare the absorption cross sections under strong coupling as obtained from a full calculation without approximations to those obtained within the BOA, for a range of coupling strengths E_{1ph} to the EM mode. In the bottom rows we also include the corresponding PoPES with one excitation. For the case of R6G-like molecules with small vibrational spacing in Fig. 3.4 we find that even for relatively small E_{1ph} the BOA agrees almost perfectly with the full results. However, for the anthracene-like molecule with a high-frequency vibrational mode and large offset ΔR, the BOA only agrees with the full result for relatively large values of E_{1ph}, where the Rabi splitting Ω_R (as defined by the energy difference between the two "polariton" peaks in the absorption spectrum) is appreciably larger than the vibrational frequency $\omega_v \approx 180$ meV (Fig. 3.5). As an aside, we note here that for intermediate values of the coupling strength (e.g., for $E_{1ph} = 0.002$ a.u. in Fig. 3.5b), the EM mode strongly couples with the individual vibronic subpeaks, as observed in experiments using anthracene [7, 18].

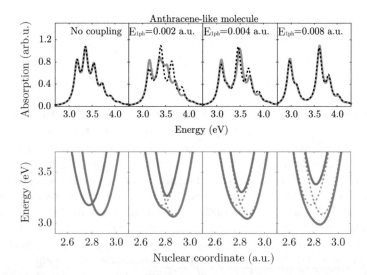

Fig. 3.5 Top row: Absorption cross sections of a single anthracenelike model molecule, calculated using the full Hamiltonian without approximation (solid green lines) and within the BOA (dashed black lines) for several values of the coupling strength. Bottom row: Corresponding single-molecule PES in the single-excitation subspace in strong coupling (solid lines) and uncoupled (dashed lines)

3.2.4 Nonadiabatic Corrections in Strong Coupling

This qualitative observation can be quantified by calculating the nonadiabtic correction $\Lambda_{\mathrm{UP,LP}}$ between the resulting PoPES. As discussed in Sect. 2.2, nonadiabatic terms become more important for small energy differences, and thus large coupling strengths are a more favorable scenario for in terms of the validity of the BOA in the absorption spectrum, as they lead to larger differences in energy. In this section we present a simple model to derive the nonadiabatic corrections induced by strong coupling without any explicit knowledge of the electronic wavefunctions. We treat the two relevant PES in the single-excitation subspace, $V_g(R) + \omega_c$ and $V_e(R)$, coupled by the term $E_{\mathrm{1ph}}\mu_{\mathrm{eg}}(R)$, where $\mu_{\mathrm{eg}}(R) = \langle e|\hat{x}|g \rangle$ is the electronic transition dipole moment between the ground and the excited states. This leads to a 2×2 Born–Oppenheimer Hamiltonian of the form

$$\hat{H}(R) = \begin{pmatrix} V_g(R) + \omega_c & E_{\mathrm{1ph}}\mu_{\mathrm{eg}}(R) \\ E_{\mathrm{1ph}}\mu_{\mathrm{eg}}(R) & V_e(R) \end{pmatrix}, \tag{3.6}$$

which can be easily diagonalized to obtain polariton eigenstates $|+\rangle = \cos\theta\,|g, 1\rangle + \sin\theta\,|e, 0\rangle$ and $|-\rangle = \sin\theta\,|g, 1\rangle - \cos\theta\,|e, 0\rangle$, where $|a, n\rangle$ is short for $|\phi_a(x; R), n\rangle$, and

$$\tan 2\theta = \frac{2h(R)}{\delta V(R)}, \tag{3.7}$$

where we defined $\delta V(R) = V_g(R) + \omega_c - V_e(R)$ and $h(R) = E_{1ph}\mu_{eg}(R)$. Using $V_{avg}(R) = \left[V_g(R) + \omega_c + V_e(R)\right]/2$, the eigenenergies are given by

$$V_\pm(R) = V_{avg}(R) \pm \frac{1}{2}\sqrt{4h^2(R) + \delta V(R)^2}. \tag{3.8}$$

With this model we can now evaluate the nonadiabatic coupling terms, which we can rewrite as $\hat{\Lambda}_{kk'} = \langle k|\,\hat{T}_n\,|k'\rangle + \langle k|\,\frac{\hat{P}}{M}\,|k'\rangle\,\hat{P}$. In order to obtain simple analytical results we can introduce a series of approximations. First, we linearize $\delta V(R) \approx a_0(R - R_c)$ around the point of intersection between the two PES, where $V_g(R_c) + \omega_c = V_e(R_c)$. Second, in the spirit of the Franck–Condon approximation, we assume that the dipole coupling is constant over the range of relevant R-values, and set $h(R) = h_0$. Following the same idea, we additionally assume that the electronic wave functions do not change significantly with R, i.e., $\frac{\partial}{\partial R}|\phi_a(x;R)\rangle \approx 0$. This implies that the change in the polaritonic eigenfunctions $|\pm\rangle$ close to the avoided crossing at R_c is mostly due to the switchover between the two uncoupled surfaces, i.e., the change in $\theta(R)$, not because of an intrinsic change of electronic state with R. With these approximations, the correction terms become

$$\langle -|\,\hat{P}\,|+\rangle = \frac{-ia_0h_0}{4h_0^2 + a_0^2(R - R_c)^2}, \tag{3.9a}$$

$$\langle -|\,\hat{P}^2\,|+\rangle = \frac{2a_0^3h_0(R - R_c)}{(4h_0^2 + a_0^2(R - R_c)^2)^2}, \tag{3.9b}$$

$$\langle \pm|\,\hat{P}^2\,|\pm\rangle = \frac{a_0^2h_0^2}{(4h_0^2 + a_0^2(R - R_c)^2)^2}, \tag{3.9c}$$

with the diagonal terms $\langle \pm|\,\hat{P}\,|\pm\rangle$ identically zero. Note that diagonal terms only correspond to energy shifts and do not induce additional transitions [2]. The nonadiabatic coupling between the polariton surfaces has a Lorentzian shape around the avoided crossing, and as expected only becomes non-negligible close to it.

As shown in Fig. 3.6, the nonadiabatic corrections obtained from this simple model agree almost perfectly with those obtained from the full numerical calculation for our anthracene-like model molecule. The only molecule-specific information entering the model are the PES of the uncoupled molecule and the dipole moment between the coupled surfaces. Specifically, the *electronic* wave functions are never used here, and their derivative as a function of the nuclear coordinates is not required. This implies that this model could be used to obtain accurate non-BO corrections that describe the transitions between potential surfaces even when the full electronic wave functions of a molecule are not available (e.g., in density-functional-theory calculations). The dynamics of the molecule could thus be fully recovered within a potential energy surface picture even when the BOA per se is not applicable.

We now exploit this model to derive a condition for which the BOA becomes a better approximation, i.e., when the nonadiabatic terms become negligible. We approximate the bare molecular potential energy surfaces as two harmonic oscilla-

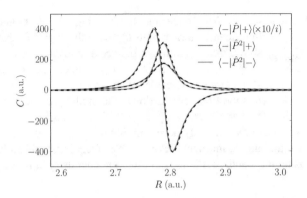

Fig. 3.6 Nonadiabatic correction terms C that couple the "lower polariton" and "upper polariton" PES for a single anthracene-like model molecule for a coupling strength of $E_{1ph} = 0.002$ a.u.. Solid colored lines: results from a full numerical calculation. Dashed black lines: results from the model Eq. (3.9). Note that while all results are given in atomic units, the units of the $\hat{P}$ and $\hat{P}^2$ terms are not identical, and thus not directly comparable

tors with the same vibrational frequency ω_v, but with an offset in energy ΔV and equilibrium position ΔR,

$$V_g(R) \approx \frac{M\omega_v^2}{2} R^2, \tag{3.10}$$

$$V_e(R) \approx \frac{M\omega_v^2}{2} (R - \Delta R)^2 + \Delta V, \tag{3.11}$$

where without loss of generality, we have chosen the origin in nuclear coordinate and energy at the minimum of $V_g(R)$. Note that this model exactly results from the common approximation of linear coupling between a single electronic excitation and a bosonic vibrational mode [19–22]. Within this model, $\delta V(R) = V_g(R) + \omega_c - V_e(R) = a_0(R - R_c)$ is already exactly linear, i.e., the linearization of the energy difference performed above is not an approximation. The constants are given by $a_0 = M\omega_v^2 \Delta R$ and $R_c = \frac{\Delta R}{2} + \frac{\Delta V - \omega_c}{a_0}$. The maximum value of $|\langle +| \frac{\hat{P}}{M} |-\rangle|$, reached at $R = R_c$, is given by $\Delta R \omega_v^2/(4h_0)$. Comparing this with the energy splitting at that point, $V_+(R_c) - V_-(R_c) = 2h_0$, gives the condition that $\Delta R \omega_v^2/(8h_0^2)$ must be small compared to the momentum of the respective nuclear wavefunction (due to the additional $\hat{P}$ operating on the nuclear wave function). The off-diagonal terms $\langle -| \frac{\hat{P}^2}{2M} |+\rangle$ reach a maximum value (again relative to the detuning) of $M\Delta R^2 \omega_v^4/(25\sqrt{5}h_0^3)$ at $R = R_c + h_0/(M\Delta R\omega_v^2)$.

By analyzing the analytical conditions obtained we find that the model molecules present two opposite cases for the applicability of the BOA: our R6G-like molecule has a relatively small vibrational spacing $\omega_v \approx 70$ meV and small electron-phonon coupling, $\Delta R \approx 0.018$ a.u., while our anthracene-like model molecule has a large vibrational spacing $\omega_v \approx 180$ meV and large electron-phonon coupling, $\Delta R \approx$

0.092 a.u.. We note that in many experiments involving organic molecules, $\Omega_R \gtrsim$ 500 meV [6, 8] is significantly larger than typical vibrational frequencies $\omega_v \lesssim$ 200 meV [23]. This shows that the intuitive picture of nuclear dynamics unfolding on uncoupled Born–Oppenheimer potential energy surfaces can often be applied to understand the modification of molecular chemistry induced by strong coupling. Additionally, even when the BOA breaks down, the model presented in here can be used to obtain the nonadiabatic coupling terms without requiring knowledge of the electronic wave functions. The only necessary input are the uncoupled PES and the associated transition dipole moments. Even for relatively large molecules, these can be obtained using the standard methods of quantum chemistry or density functional theory.

3.3 Two Molecules

In this section we will study the case for two molecules. As discussed in Sect. 2.3, the inclusion of more than one molecule leads to collective strong coupling, where the linear combination of N emitters so-called bright state couples to the single photonic mode, while $N - 1$ dark states appear. We thus expect to observe this phenomenon in the following analysis.

3.3.1 Method

We now treat the case of two model molecules, which can still be solved exactly within our approach, but which displays many of the effects of many-molecule strong coupling. We extend the Hamiltonian of Eq. (3.4) for two molecules coupled to a single photonic mode and with no direct dipole–dipole interaction, similarly as in the TC model (see Sect. 2.3). Then the Hamiltonian for two molecules is

$$\hat{H}^{2m}_{c-mol} = \omega_c \hat{a}^\dagger \hat{a} + \sum_{j=1,2} \left(\hat{H}^{(j)}_{mol} + E_{1ph} \hat{\mu}^{(j)} (\hat{a}^\dagger + \hat{a}) \right), \qquad (3.12)$$

where the superscripts j indicate the molecule on which the operator acts. Directly diagonalizing this Hamiltonian in the "raw" basis $\{x_1, R_1, x_2, R_2, n\}$ is already a formidable computational task for typical grid sizes. We thus calculate the full solution by first diagonalizing the single-molecule Hamiltonian, $\hat{H}_{mol} = \sum_k E_k |k\rangle \langle k|$, and including only a relevant subset of eigenstates $\{k\}$ for each molecule in the total basis $\{k_1, k_2, n\}$. The number of necessary eigenstates to obtain completely converged results is quite small ($\approx$30 per molecule). However, this approach only provides limited insight into the dynamics of the strongly coupled system, especially regarding nuclear motion.

We thus again apply the Born–Oppenheimer approximation by separating the nuclear kinetic energy terms and diagonalizing the remaining Hamiltonian parametrically as a function of R_1 and R_2. Similar to above, instead of working in the $\{x_1, x_2, n\}$ basis for each combination (R_1, R_2), we prediagonalize the single-molecule electronic Hamiltonian $\hat{H}_e(x; R) = \sum_k V_k(R) |k(R)\rangle \langle k(R)|$, where (for the cases discussed here) the sum only has to include the ground and first excited states to achieve convergence, $k \in \{g, e\}$. If we additionally allow at most one photon in the system, $n \in \{0, 1\}$, we obtain an 8×8 Hamiltonian for each combination of nuclear coordinates R_1, R_2.

The electronic Hamiltonian consists of all possible combinations of electronic states V_g, V_e of the two molecules with 0 or 1 photons. A further simplification is achieved by taking into account that the Hamiltonian conserves parity[2] $\Pi = (-1)^{\pi_1 + \pi_2 + n}$, with π_j the parity of the state of molecule j (even or odd). For large coupling E_{1ph}, this separation by parity avoids some accidental degeneracies between uncoupled PES and thus improves the BOA. We now obtain two independent 4×4 Hamiltonians,

$$\hat{H}_{\text{even}}(R_1, R_2) = \begin{pmatrix} V_{gg0} & E_{1ph}\mu^{(1)} & E_{1ph}\mu^{(2)} & 0 \\ E_{1ph}\mu^{(1)} & V_{eg1} & 0 & E_{1ph}\mu^{(1)} \\ E_{1ph}\mu^{(2)} & 0 & V_{ge1} & E_{1ph}\mu^{(2)} \\ 0 & E_{1ph}\mu^{(1)} & E_{1ph}\mu^{(2)} & V_{ee0} \end{pmatrix}, \tag{3.13a}$$

$$\hat{H}_{\text{odd}}(R_1, R_2) = \begin{pmatrix} V_{gg1} & E_{1ph}\mu^{(1)} & E_{1ph}\mu^{(2)} & 0 \\ E_{1ph}\mu^{(1)} & V_{eg0} & 0 & E_{1ph}\mu^{(1)} \\ E_{1ph}\mu^{(2)} & 0 & V_{ge0} & E_{1ph}\mu^{(2)} \\ 0 & E_{1ph}\mu^{(1)} & E_{1ph}\mu^{(2)} & V_{ee1} \end{pmatrix}, \tag{3.13b}$$

where the uncoupled PES are represented by the compact notation $V_{\alpha\beta n} = V_\alpha(R_1) + V_\beta(R_2) + n\omega_c$, and the single-molecule dipole transition moment between the ground and first excited state is denoted by $\mu^{(j)} = \langle \phi_g(R_j) | \hat{\mu} | \phi_e(R_j) \rangle$. Diagonalizing these Hamiltonians for each (R_1, R_2) results in a set of two-dimensional PoPES. In Fig. 3.7, we show the three surfaces in the single-excitation subspace, corresponding to the three lowest states of Eq. (3.13b). For zero molecule-photon coupling ($E_{1ph} = 0$, Fig. 3.7a), there are now a number of one-dimensional seams where the three PES cross. When the molecule-photon coupling is turned on, these crossings again turn into avoided crossings, as shown in panels (b) and (c) for two different coupling strengths E_{1ph}. Following the natural convention discussed in previous chapters, we label the three PoPES in order of energy as the "lower polaritonic PES", the "dark-state PES", and the "upper polaritonic PES".

---

[2] We note that this is only true because we do not have permanent dipole moments in our model, which couple states of the form $|k, k, 0\rangle$ and $|k, k, 1\rangle$.

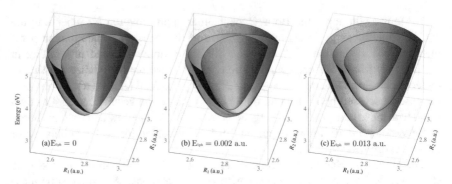

Fig. 3.7 **a** Uncoupled potential energy surfaces of two anthracene-like molecules in the singly excited subspace: $V_{ege0}(R_1, R_2)$ (orange), $V_{ege0}(R_1, R_2)$ (blue), and $V_{gg1}(R_1, R_2)$ (green). **b** Coupled PES for $E_{1ph} = 0.002$ a.u. and **c** $E_{1ph} = 0.013$ a.u., corresponding to the lower polariton (orange), dark state (blue), and upper polariton (green). For clarity, only parts where $R_1 < R_2$ are shown (note that the system is symmetric under the exchange $R_1 \leftrightarrow R_2$)

We first address the applicability of the BOA, which breaks down when two PoPES are close in energy, for the case of two molecules. Within the single-excitation subspace (which determines the linear properties of the system, such as absorption), there are now a range of (avoided) crossings. They occur when (i) all three surfaces approach each other $V_{gg1} \approx V_{ge0} \approx V_{eg0}$, (ii) the photonically excited PES is close to only one of the electronically excited PES, $V_{gg1} \approx V_{ge0}$ or $V_{gg1} \approx V_{eg0}$, or (iii) only the two electronically excited states cross, $V_{ge0} \approx V_{eg0}$. Case (i) corresponds to the TC model at zero detuning, giving the two polaritonic PES at energy shifts of $\pm\Omega_R/2$, and an additional dark state that is unshifted from the bare-molecule case. The BOA in this region is thus valid for similar conditions as in the single-molecule case, although the PES separation is reduced by half due to the additional dark-state surface. Case (ii) corresponds exactly to the single-molecule case, with the second molecule acting as a "spectator" that only induces additional energy shifts. The BOA should thus again be valid for similar conditions as with a single-molecule, albeit with the coupling reduced by $1/\sqrt{2}$ for a fixed total Rabi splitting. Finally, case (iii) presents the biggest challenge, as the two electronically excited PES, V_{eg0} and V_{ge0}, are not directly coupled, but only split indirectly through coupling to the photonically excited surface V_{gg1}. The splitting between the two surfaces is thus small for large detuning, $\Delta V \approx (E_{1ph}\mu)^2/4(V_{gg1} - V_{eg0})$. This is clearly observed in Fig. 3.7b along the line $R_1 = R_2$, where the dark state PES almost touches the upper PoPES for small Rs and the lower PoPES for large Rs.

3.3.2 Absorption

The discussion above implies that for almost any coupling strength, there will be regions in the nuclear configuration space R_1, R_2 where the BOA breaks down. However, not all parts of the PES are visited by the nuclei during a given physical process. To explicitly check the BOA in the subspace relevant for polaritonic physics, in Fig. 3.8 we thus again compare the absorption obtained within the BOA with that computed by a full diagonalization of the Hamiltonian Eq. (3.12). Compared to the single-molecule case, many more molecular levels are present in the system, leading to small changes in the absorption spectra compared to the single-molecule case. In order to properly compare the results, we use the same Rabi splitting. In the picture of PoPES, we cannot use the usual definition of $\Omega_R = E_+ - E_-$ of the TC model. In the following we therefore define it as the corresponding separation in equilibrium (minimum of the bare-molecule ground-state surface $V_g(R)$) and for zero detuning, which for a collection of many molecules is

$$\Omega_R = 2\sqrt{N} E_{1ph} \mu(R_{eq}), \tag{3.14}$$

where we can tune the single-photon electric field amplitude to achieve the desired Rabi splitting in our calculations. We therefore take into account the $\sqrt{N}$ scaling of the total Rabi frequency and reduce the coupling strengths by $\sqrt{2}$ to produce the same total splitting in the case of two molecules.

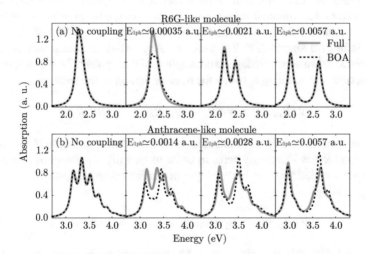

Fig. 3.8 Absorption cross section of two molecules driven coherently, calculated using the full Hamiltonian without approximation (solid green lines) and within the BOA (dashed black lines). Results are shown for the **a** R6G-like and **b** anthracene-like model molecules, for several values of the coupling strength E_{1ph}. The values of E_{1ph} are scaled by $1/\sqrt{2}$ with respect to the single-molecule case (Figs. 3.4a and 3.5a) in order to obtain the same total Rabi frequency Ω_R

The BOA is shown to again give good results for large enough coupling, but the minimum coupling required is increased compared to that for a single molecule. In the common case of slow nuclear motion, as for our R6G-like model in Fig. 3.8a, the BOA already is valid for relatively small Rabi splitting of $\Omega_R \approx 250$ meV. However, in the anthracene-like case of very fast vibrational motion, Fig. 3.8b, the BOA still does not give perfect agreement with the full model for $E_{1ph} = 0.0057$ a.u. ($\Omega_R \approx 600$ meV), and agreement is only reached at roughly twice that value.

In Fig. 3.7 is clear that the new dark state PES is the main source of nonadiabatic terms that lead to the break down of the BOA. The coupling around some regions in nuclear coordinate space can be proportional to the single-molecule coupling, which, at fixed Rabi splitting, goes down when the number of molecules is increased. As discussed in the previous section, a smaller coupling strength leads to larger nonadiabatic terms. While this would appear to affect the validity of the BOA close to these configurations, we emphasize that these corrections can be computed using standard methods of quantum chemistry or density functional theory. The picture of PoPES is thus not hindered by increasing the number of molecules, as it provides an intuitive and powerful description of the system, from which nonadiabatic corrections can be easily computed.

3.3.3   Nuclear Correlation

Having established the validity of the BOA for many relevant cases and Rabi splittings comparable to experimental values, we now discuss the implications of collective strong coupling for the internal molecular (nuclear) dynamics. The BOA provides a straightforward approach to this problem. Any two-dimensional surface can be decomposed into a sum of independent single-molecule potentials, plus a remainder that describes the coupling between the nuclear motion of the molecules,

$$V(R_1, R_2) = V_1(R_1) + V_2(R_2) + V_{12}(R_1, R_2). \tag{3.15}$$

The nuclear motion of two molecules is independent if and only if the coupled part $V_{12}(R_1, R_2)$ is identically zero. In order to quantify this coupling, we expand each of the coupled surfaces in the single-excitation subspace around its minimum (R_1^0, R_2^0), giving

$$V(R_1, R_2) \approx V_0 + \alpha\, \delta R_1^2 + \alpha\, \delta R_2^2 + \beta\, \delta R_1 \delta R_2, \tag{3.16}$$

with $V_0 = V(R_1^0, R_2^0)$ and $\delta R_i = R_i - R_i^0$. Note that due to symmetry under the exchange $R_1 \leftrightarrow R_2$, the prefactor α is the same for δR_1^2 and δR_2^2. As can be seen in Fig. 3.9a, both the polariton and even the dark state PES show significant coupling of the nuclear degrees of freedom, with values of β/α on the order of a few percent for values of $E_{1ph} \lesssim 0.01$ a.u. giving Rabi splittings of $\lesssim 1$ eV (see Fig. 3.8). Interestingly, the coupling is much larger for the lower PoPES than for either the upper PoPES

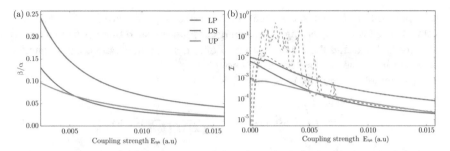

Fig. 3.9 **a** Coupling between nuclear motion in different molecules for the lower (LP) and upper polariton (UP) and dark-state (DS) PES. Results are shown as the ratio β/α between the prefactors of the offdiagonal R_1R_2 and diagonal R_i^2 terms in Eq. (3.16), for the R6G-like model molecule. **b** Mutual information in the nuclear probability distribution for two different states. The solid lines correspond to the vibrational ground states of the coupled PES as predicted by Eq. (3.19) using the values of β/α shown in (**a**), while the dashed lines correspond to the steady state under driving of a single molecule as defined in Eq. (3.17)

or the dark state PES, and decreases with increasing E_{1ph} for all three surfaces. We therefore conclude that even dark states that have negligible mixing with photonic modes are affected by strong coupling, in the sense that the nuclear degrees of freedom of separate molecules behave like coupled harmonic oscillators, and their motion becomes correlated. This implies that, e.g., local excitation of nuclear motion within one molecule could affect the nuclear motion in another, spatially separated molecule, even when no photon is ever present in the EM mode of the system.

Note that the BOA results predict monotonously increasing correlation for arbitrarily small (but non-zero) values of E_{1ph}. This again shows that the BOA is not correct in the limit of small coupling $E_{1ph} \rightarrow 0$, where the correlation should also go to zero as the molecules are completely uncoupled. We thus start Fig. 3.9a at $E_{1ph} = 0.002$ a.u., for which the BOA already produces good agreement with the full result in the absorption cross section (see Fig. 3.8), and note that our results indicate that there is a maximum of correlation in the nuclear motion at intermediate coupling strengths.

In order to verify these results outside the BOA, we calculate the mutual information in the nuclear probability distribution both for the harmonic expansion of Eq. (3.16) within the BOA and under external driving of a single molecule. From first-order perturbation theory, the driven steady state is given by

$$\left| \psi_1^{dr}(\omega) \right\rangle = \frac{1}{\hat{H} - \omega_0 - \omega - i\epsilon} \hat{\mu}_1 \left| \psi_0 \right\rangle \tag{3.17}$$

which we solve using the full Hamiltonian without approximations. We again use a non-zero ϵ to artificially represent losses in the system (for the results below, we choose $\epsilon = 2.5$ meV, corresponding to an effective decay rate of 5 meV). While the eigenstates of the Hamiltonian split into quasidegenerate symmetric and antisymmetric superpositions (which show large correlation) for any non-zero E_{1ph}, the non-zero value of ϵ leads to a smearing of the energy resolution, such that the degeneracy is effectively lifted and the superposition of only a single molecule being excited is observed in the steady state for small enough E_{1ph}.

The mutual information is calculated as [24]

$$\mathcal{I} = \iint P(R_1, R_2) \log_2 \frac{P(R_1, R_2)}{P(R_1)P(R_2)} dR_1 dR_2, \qquad (3.18)$$

with $P(R_1, R_2)$ being the joint nuclear probability distribution for either the driven steady-state wave function $\left| \psi_1^{dr}(\omega) \right\rangle$ or the ground state of two coupled harmonic oscillators. For the latter, $\mathcal{I}$ can be analytically calculated as

$$\mathcal{I}_0 = \log_2 \frac{\sqrt{2 - \beta/\alpha} + \sqrt{\beta/\alpha + 2}}{2\sqrt[4]{4 - \beta^2/\alpha^2}}. \qquad (3.19)$$

In order to compare with the predictions obtained from the ratio β/α for the PES, we choose driving frequencies ω equal to the vibrational ground state energies of each surface. The dashed lines in Fig. 3.9b show that, as could be expected, at zero coupling ($E_{1ph} = 0$) there is no correlation under driving of a single molecule. As E_{1ph} increases, the mutual information quickly increases and actually becomes significantly larger than the BO ground-state values for the DS and upper PoPES. In this region, there are a series of avoided crossings, and the results are expected to depend strongly on the correct description of decay and dephasing, which we only treat phenomenologically. For larger E_{1ph}, where the BOA becomes valid, the mutual information in the driven steady state agrees very well with the mutual information as predicted from the coupling β/α in the Taylor expansion of the PES. Interestingly, the agreement between the full calculation and the BO result for the LP PES is very good even at relatively low coupling strengths. This is a consequence of the fact that the LP ground state is well-isolated in energy, while the DS and UP surfaces are not. We believe that this property is also related to the experimentally observed fast nonradiative decay of upper polariton states [6, 13, 25, 26], which can take place efficiently close to avoided crossings of the PES (where the BOA breaks down).

3.4   Conclusions

In this chapter we studied in detail how strong coupling can influence the internal structure of organic molecules, and the limitations of the usual Born–Oppenheimer picture. We show under which conditions the molecular properties under strong coupling can be understood by the modification of the potential energy surfaces determining nuclear dynamics under the Born–Oppenheimer approximation. In particular, we found that in many cases of experimental interest where the Rabi splitting is large, the BOA is applicable and provides an intuitive picture of the strongly coupled dynamics. However, we have also shown that for molecules with fast vibrational modes and large phonon-exciton couplings, the BOA can break down and nonadiabatic corrections are required in order to fully describe the energy landscape. We furthermore demonstrated that the nonadiabatic coupling terms between PES in this case are dominantly due to the change of character between light and matter excitations which can be obtained from simple few-level models without requiring knowledge of the electronic wavefunctions.

In addition, we show that under collective strong coupling involving more than one molecule, the nuclear dynamics of the molecules in electronic "dark states" that are only weakly coupled to the photonic mode are nonetheless affected by the formation of strong coupling. In particular, we find that the dark state PES describes coupling between the nuclear degrees of freedom of the different molecules.

These results validate the use of the Born–Oppenheimer approximation in molecular polaritonics and thus lay the groundwork for the following chapter, where a more general theory of polaritonic chemistry is developed. This theory is based on the concept that we have introduced here of polaritonic potential energy surfaces, which extends the usual PES of chemistry to hybrid light–matter systems participating a collection of molecules.

References

1. Born M, Oppenheimer R (1927) Zur Quantentheorie der Molekeln. Annalen der Physik 389:457
2. Tully JC (2000) Perspective on "Zur Quantentheorie der Molekeln". Theor Chem Acc Theory Comput Model (Theoretica Chimica Acta) 103:173
3. Chikkaraddy R, de Nijs B, Benz F, Barrow SJ, Scherman OA, Rosta E, Demetriadou A, Fox P, Hess O, Baumberg JJ (2016) Single-molecule strong coupling at room temperature in plasmonic nanocavities. Nature 535:127
4. Zengin G, Wersäll M, Nilsson S, Antosiewicz TJ, Käll M, Shegai T (2015) Realizing strong light-matter interactions between single-nanoparticle plasmons and molecular excitons at ambient conditions. Phys Rev Lett 114:157401
5. Lidzey DG, Bradley DDC, Skolnick MS, Virgili T, Walker S, Whittaker DM (1998) Strong exciton-photon coupling in an organic semiconductor microcavity. Nature 395:53
6. Schwartz T, Hutchison JA, Genet C, Ebbesen TW (2011) Reversible switching of ultrastrong light-molecule coupling. Phys Rev Lett 106:196405
7. Kéna-Cohen S, Davanço M, Forrest SR (2008) Strong exciton-photon coupling in an organic single crystal microcavity. Phys Rev Lett 101:116401
8. Kéna-Cohen S, Maier SA, Bradley DD (2013) Ultrastrongly coupled exciton-polaritons in metal-clad organic semiconductor microcavities. Adv Opt Mater 1:827
9. Houdré R, Stanley RP, Ilegems M (1996) Vacuum-field Rabi splitting in the presence of inhomogeneous broadening: resolution of a homogeneous linewidth in an inhomogeneously broadened system. Phys Rev A At Mol Opt Phys 53:2711
10. Agranovich V, Gartstein YN, Litinskaya M (2011) Hybrid resonant organic-inorganic nanostructures for optoelectronic applications. Chem Rev 111:5179
11. Michetti P, Mazza L, La Rocca GC (2015) Strongly coupled organic microcavities. In: Zhao YS (ed) Organic nanophotonics, nano-optics and nanophotonics, vol 39. Springer, Berlin, Heidelberg
12. Gonzalez-Ballestero C, Feist J, Gonzalo Badía E, Moreno E, Garcia-Vidal FJ (2016) Uncoupled dark states can inherit polaritonic properties. Phys Rev Lett 117:156402

13. Sáez-Blázquez R, Feist J, Fernández-Domínguez A, García-Vidal F (2018) Organic polaritons enable local vibrations to drive long-range energy transfer. Phys Rev B 97:241407
14. Hakala TK, Toppari JJ, Kuzyk A, Pettersson M, Tikkanen H, Kunttu H, Törmä P (2009) Vacuum Rabi splitting and strong-coupling dynamics for surface-plasmon polaritons and rhodamine 6G molecules. Phys Rev Lett 103:1
15. Rodriguez SRK, Feist J, Verschuuren MA, Garcia Vidal FJ, Gómez Rivas J (2013) Thermalization and cooling of plasmon-exciton polaritons: towards quantum condensation. Phys Rev Lett 111
16. May V, Kühn O (2011) Charge and energy transfer dynamics in molecular systems. Wiley-VCH Verlag GmbH & Co. KGaA, Weinheim, Germany
17. Ciuti C, Carusotto I (2006) Input-output theory of cavities in the ultrastrong coupling regime: the case of time-independent cavity parameters. Phys Rev A At Mol Opt Phys 74
18. Kéna-Cohen S, Forrest SR (2010) Room-temperature polariton lasing in an organic single-crystal microcavity. Nat Photonics 4:371
19. Leggett AJ, Chakravarty S, Dorsey AT, Fisher MP, Garg A, Zwerger W (1987) Dynamics of the dissipative two-state system. Rev Modern Phys 59:1
20. Coalson RD, Evans DG, Nitzan A (1994) A nonequilibrium golden rule formula for electronic state populations in nonadiabatically coupled systems. J Chem Phys 101:436
21. Spano FC (2015) Optical microcavities enhance the exciton coherence length and eliminate vibronic coupling in J-aggregates. J Chem Phys 142
22. Herrera F, Spano FC (2016) Cavity-controlled chemistry in molecular ensembles. Phys Rev Lett 116:238301
23. Shimanouchi T (1972) Tables of molecular vibrational frequencies. Consolidated volume I. NSRDS-NBS 39:161p
24. Gray RM (1990) Entropy and information theory, 1st edn. Springer
25. Ribeiro RF, Martínez-Martínez LA, Du M, Campos-Gonzalez-Angulo J, Yuen-Zhou J (2018) Polariton chemistry: controlling molecular dynamics with optical cavities. Chem Sci 9:6325
26. Pino JD, Feist J, Garcia-Vidal FJ (2015) Quantum theory of collective strong coupling of molecular vibrations with a microcavity mode. New J Phys 17:53040

Chapter 4
Theory of Polaritonic Chemistry

4.1 Introduction

In the previous chapter we embraced the complexity of organic molecules by including arbitrary PES in our description of strong coupling with the aim of building a
general theory of polaritonic chemistry. In this chapter we generalize these results
and analyze the approach of PoPES. We study the general light–matter Hamiltonian
from the point of view presented in Chap. 3, i.e., by separating the electronic and
photonic DoF from the nuclear coordinates. In Sect. 4.2 we explicitly analyze this in a
complete and general way, also presenting a conceptual molecular energy landscape
that presents some kind of excited-state process that can be strongly influenced in
strong coupling. Then, in Sect. 4.3 we present the potential of the PoPES picture for
describing collective phenomena. We show how we can use the spin operators used
in the Tavis–Cummings model (see Sect. 2.3) to understand such a complex system.
The ensemble of N molecules is formally identical to a single "supermolecule" that
encompass the internal DoF of *all* molecules. This immediately leads to novel phenomena such as the collective protection effect and collective conical intersections,
both discussed in detail in this section.

4.2 Polaritonic Potential Energy Surfaces

In order to theoretically describe phenomena in polaritonic chemistry we need to
extend the chemistry formalism discussed in Sect. 2.2 to a collection of N molecules
coupled to one or several quantized light modes. The total Hamiltonian is given by

$$\hat{H}_{tot} = \sum_{i}^{N} \hat{T}_{n}^{(i)} + \sum_{i}^{N} \hat{H}_{e}^{(i)} + \hat{H}_{EM} + \sum_{i}^{N} \hat{H}_{int}^{(i)}, \tag{4.1}$$

© The Editor(s) (if applicable) and The Author(s), under exclusive license
to Springer Nature Switzerland AG 2020
J. Galego Pascual, *Polaritonic Chemistry*, Springer Theses,
https://doi.org/10.1007/978-3-030-48698-3_4

where the kinetic energy operator and the electronic Hamiltonians of all molecules have been explicitly taken into account. Note that in Eq. (4.1) we directly disregarded the direct dipole–dipole interaction between molecules. Additionally, note that we neglect the dipole self-interaction term as discussed in Sect. 2.3, restricting the generality of this Hamiltonian to quasistatic cavities [1] or to coupling strengths outside the ultrastrong and deep coupling regimes. Finally, it should be noted that the Hamiltonian in Eq. (4.1) is also the starting point of more reduced models such as the TC Hamiltonian, or the Holstein–Tavis–Cummings model [2].

The picture of PoPES is inspired by the formal similarities between Eq. (4.1) and the standard molecular Hamiltonian discussed in Sect. 2.2

$$\hat{H}_{mol} = \hat{T}_n + \hat{H}_e(\mathbf{R}).$$ (4.2)

In the previous chapter we studied the adiabatic separation of nuclear and electron–photon energies from first principles. We now explicitly write the general "electronic–photonic" Hamiltonian

$$\hat{H}_{e-ph}(\mathbf{q}) = \hat{H}_{tot} - \sum_i^N \hat{T}_n^{(i)},$$ (4.3)

where $\mathbf{q} = (\mathbf{R}_1, \mathbf{R}_2, \ldots, \mathbf{R}_N)$ is the vector describing all nuclear coordinates of all molecules. For future reference, let us now write this Hamiltonian explicitly for one single photonic mode:

$$\hat{H}_{e-ph}(\mathbf{q}) = \omega_c \hat{a}^\dagger \hat{a} + \sum_i^N \left(\hat{H}_e^{(i)}(\mathbf{R}_i) + \mathbf{E}_{1ph,i} \cdot \hat{\boldsymbol{\mu}}_i(\mathbf{R}_i)(\hat{a}^\dagger + \hat{a}) \right),$$ (4.4)

where the exciton–photon interaction is determined by the single-photon electric field amplitude $\mathbf{E}_{1ph,i}$, which may be different from one molecule to another. Analogously as for a single molecular Hamiltonian, the diagonalization of $\hat{H}_{e-ph}(\mathbf{q})$ yields an adiabatic basis of hybrid electron–photon states $\{\Phi_k(\mathbf{q})\}$ with the corresponding PoPES $V_k(\mathbf{q})$.

It should be noted that this new electronic–photonic basis used in the Schrödinger equation leads to the usual set of differential equations analogous to Eq. (2.29), with new nonadiabatic terms. In this picture we can again perform the BOA and neglect these terms. However, these nonadiabatic corrections $\Lambda_{kk'}$ in the new adiabatic basis of polaritonic states now contain both the original bare-molecule nonadiabatic couplings (appropriately transformed to the polaritonic basis) and the light–matter induced nonadiabatic couplings, as discussed in Chap. 3.

The electronic–photonic Hamiltonian can be diagonalized in two stages, first by doing the appropriate adiabatic separation in each individual molecule (i.e., we can express $\hat{H}_e^{(i)}(\mathbf{R}_i)$ in Eq. (4.1) already in the adiabatic single-molecule electronic basis), and then selecting the relevant subset of bare molecular states coupled to the photonic modes. For instance, as discussed in Sect. 2.3, for moderate values of

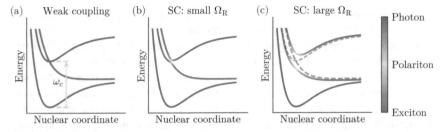

Fig. 4.1 Conceptual potential energy surfaces for a single molecule describing a typical bound and dissociative energy landscape coupled to a light mode in **a** weak coupling and **b, c** strong coupling for different coupling strengths. The color represents the photonic fraction of the state from purely excitonic (orange) through polaritonic (light gray) to purely photonic (purple). Reprinted with permission from [3]. Copyright 2018 American Chemical Society

the coupling strength the TC model conserves the number of excitations. This is also true here, which means that for a single photonic mode, we can restrict the number of electronic states to the ones close in energy to the photon frequency. This formulation becomes a compelling tool, as it is straightforward to make an interface with existing quantum chemistry methods, which can be used to calculate the bare-molecule structure for the desired electronic states at each configuration $\mathbf{R}$ for each molecule separately.

The light–matter coupling can be readily treated with a small Hamiltonian involving only a few relevant states per molecule. In the following we will study the general properties of the PoPESs by treating a minimal model of a single molecule with two electronic states and one nuclear DoF, i.e., the bare molecule is characterized by a ground and excited PES, $V_g(q)$ and $V_e(q)$ respectively. In Fig. 4.1a we present a typical bound and dissociative energy landscape of diatomic molecules for a conceptual molecule (see blue and orange lines in Fig. 4.1a), with an additional purple surface representing the ground-state molecule plus one photon, of energy $V_g(q) + \omega_c$. The coupling to the electronic part is introduced within the RWA (see the discussion of the Tavis–Cummings model in Sect. 2.3), such that the total number of electronic and photonic excitations is conserved, but leads to incorrect results when treating the ultrastrong-coupling regime. The relevant electronic–photonic Hamiltonian becomes

$$\hat{H}_{e-ph}(q) = V_g(q) + V'_e(q)\hat{\sigma}^\dagger\hat{\sigma} + \omega_c\hat{a}^\dagger\hat{a} + \mathbf{E}_{1ph}\cdot\boldsymbol{\mu}_{eg}(q)\left(\hat{a}^\dagger\hat{\sigma} + \hat{\sigma}^\dagger\hat{a}\right), \quad (4.5)$$

where $V'_e(q) = V_e(q) - V_g(q)$ is the position-resolved energy difference between ground- and excited-state PESs, $\hat{\sigma} = |g\rangle\langle e|$ is the molecular electronic transition operator, ω_c the photon frequency, and $\hat{a}$ the bosonic ladder operator associated to the photon. The exciton–photon interaction is determined by the single-photon electric field amplitude $\mathbf{E}_{1ph}$ and the configuration-dependent electronic transition dipole moment $\boldsymbol{\mu}_{eg}(q)$ between ground and excited states. It should be noted that the permanent dipole moments are also omitted, and thus Eq. (4.5) is very similar to the Jaynes–Cummings Hamiltonian. Diagonalization of $\hat{H}_{\text{e−ph}}(q)$ yields the set of adiabatic PoPES of the strongly-coupled system.

When the coupling is negligible (Fig. 4.1a), two clearly distinguishable PES exist in the single-excitation subspace: the molecular exciton, characterized by the excitonic PES (orange line), and the state corresponding to a single cavity photon, with the molecule in its ground state (thus the PES inherits the ground-state like behavior with a shift up the photon energy, ω_c, purple line). Throughout most of the plots in this thesis that feature PoPES, we codify the mixed light–matter character in a color scale that measures the photon component $n_{\text{ph}} = \langle \hat{a}^\dagger \hat{a} \rangle$, spanning from orange (bare exciton) through light gray (polariton) to purple (bare photon). We can see this in Fig. 4.1b, c, where the Rabi splitting is increased, thus showing a modification in the energy landscape, as discussed in the previous chapter.

Let us now address the role of dissipative processes in polaritonic chemistry. Organic molecules coupled with light modes present different mechanisms of decay and dephasing. The decay processes associated with the cavity are often very fast, showing in most current experiments typical lifetimes on the order of tens to hundreds of femtoseconds [4–8], to tens of picoseconds for the case of high-Q cavities such as Fabry–Perot or photonic crystal cavities [9, 10]. Regarding the molecular part, the dissipative processes emerge from inter- and intra-molecular vibrational relaxation, i.e., interactions with all the DoF of the molecule itself and the molecular environment. These processes can in principle be described within the framework of PoPES, since the phenomena of nuclear relaxation, dephasing, and nonradiative decays take place on the PES of the molecule, and can thus be included. As an example, in [11] the molecule is treated fully with all relevant nuclear degrees of freedom and the solvent molecules are represented through molecular mechanics. An additional decay channel that molecules present is the free space radiative decay, determined by Fermi's golden rule (see Sect. 2.2), which leads to typical lifetimes of the order of nanoseconds. Throughout this thesis, we neglect dissipative processes in our theoretical treatment, and only discuss their overall effects on polaritonic chemistry when studying particular scenarios (see e.g., Chap. 6). However, a more detailed treatment of dissipation will certainly be beneficial to a complete understanding of the experimental implementations of polaritonic chemistry.

4.3 Collective Phenomena: The Supermolecule

Here we analyze the effects found when strong coupling is achieved through collective coupling, i.e., the coherent interaction of many molecules with the same light mode. As already discussed, this leads to an electronic–photonic Hamiltonian that depends parametrically on $\mathbf{q} = (\mathbf{R}_1, \mathbf{R}_2, \dots, \mathbf{R}_N)$, i.e., on the nuclear degrees of freedom of all involved molecules. This property is inherited by the resulting PoPES after diagonalization of $\hat{H}_{\text{e−ph}}(\mathbf{q})$, implying that the effective indirect intermolecular interaction through the photonic mode could lead to novel correlations between the nuclei of different molecules. Indeed, this has been analyzed in detail in Sect. 3.3

for the case of two molecules, presenting remarkable correlations even in the dark states. We can thus understand collective strong coupling in organic molecules as leading to the formation of a "supermolecule" spanning all coupled molecules. As we discuss in this section, this offers a range of new phenomena that further enhance our ability to control the dynamics and chemistry of a molecular ensemble.

Let us again assume that the molecules are described by only two electronic states, and in addition we take all molecules to be coupled equally to the photonic mode. The Hamiltonian is then a straightforward extension of Eq. (4.5) to include sums over all the molecules. The associated Hilbert space of such Hamiltonian becomes rapidly unmanageable. In order to treat such a large number of states in the standard TC model we define collective operators (see Sect. 2.3). This was possible because the emitters where considered to have the same energy. However, in the picture of PoPES each molecule has a unique configuration with its associated energy. Nevertheless, we can focus our study to cuts of the Hilbert space where groups of molecules share the same configuration, thus making those sets of molecules indistinguishable. For each of these groups we introduce the different collective spin operators

$$\hat{S}_{\alpha}^{\dagger} = \sum_{i_{\alpha}=1}^{N_{\alpha}} \hat{\sigma}_{i_{\alpha}}^{\dagger} \quad \text{and} \quad \hat{S}_{\alpha} = \sum_{i_{\alpha}=1}^{N_{\alpha}} \hat{\sigma}_{i_{\alpha}}, \tag{4.6}$$

where $\hat{\sigma}_i$ is the usual single-emitter operator used in Eq. (4.5) and α labels groups of molecules with the same configuration $\mathbf{R}_{i_{\alpha}} \equiv \mathbf{R}_{\alpha}$, such that $\sum_{\alpha} N_{\alpha} = N$. These correspond to spin-$N_{\alpha}/2$ operators so for a group with $N_{\alpha} = 1$ we recover the single-emitter operator $\hat{\sigma}_i$ [12]. By including sums over the groups of "identical" molecules we get the Hamiltonian

$$\hat{H}_{e-ph}(\mathbf{q}) = V_G(\mathbf{q}) + \sum_{\alpha} V_e'(\mathbf{R}_{\alpha})\hat{n}_{\alpha} + \omega_c \hat{a}^{\dagger}\hat{a} + \mathbf{E}_{1ph} \cdot \sum_{\alpha} \boldsymbol{\mu}_{eg}(\mathbf{R}_{\alpha}) \left(\hat{a}^{\dagger}\hat{S}_{\alpha} + \hat{a}\hat{S}_{\alpha}^{\dagger} \right)$$

$$\tag{4.7}$$

where $V_G(\mathbf{q}) = \sum_i^N V_g(\mathbf{R}_i)$ is the overall ground state of the system, and $\hat{n}_{\alpha} = \hat{S}_{\alpha}^z + N_{\alpha}/2$ is the excitation number operator for the group α (with $\hat{S}_{\alpha}^z = \sum_{i_{\alpha}=1}^{N_{\alpha}} \hat{\sigma}_{i_{\alpha}}^z$ the z-component of the collective spin operator). Since the Hamiltonian now contains only collective molecular operators, the electronic–photonic states can be expressed in the collective spin basis $\hat{n}_{\alpha}|n_{\alpha}\rangle = n_{\alpha}|n_{\alpha}\rangle$. Here it is more convenient to use $n_{\alpha} = m_{\alpha} + N_{\alpha}/2$ as the quantum number, instead of the usual m_{α}, which is the z-component of the spin, since n_{α} relates to a relevant physical quantity in our system: the excitation number. This reduces the size of the Hilbert space from growing with the number of molecules N, to growing with the number of groups of molecules with the same configuration. In the following it will become apparent why this can significantly reduce the size of the Hilbert space.

4.3.1 Collective Protection

We now use Eq. (4.7) to study a collection of molecules, modeled through the simple bound and dissociative PESs depicted in Fig. 4.1. Again we restrict ourselves to the single-excitation subspace and diagonalize the electronic–photonic Hamiltonian for different cuts of the full Hilbert space. We start by exploring the two-dimensional subspace determined by restricting $N - 2$ molecules to their equilibrium position q_{eq} (minimum of bound ground state PES $V_g(q)$). The corresponding lowest excited PoPES for $N = 50$ is represented in Fig. 4.2a, for a collective Rabi splitting $\Omega_R = 0.3\,\mathrm{eV}$. By inspecting this surface we can see that it does not correspond to a simple sum of independent single-particle potentials $V_{\mathrm{LP}}(q_1, q_2, \dots) = \sum_i V_i(q_i)$, which confirms our previous presumption that strong coupling implies some correlation between nuclear motion of different molecules, as we have seen in Sect. 3.3. Furthermore, the choice of this particular two-dimensional cut reveals a general feature that aids in the analysis of even higher-dimensional PoPESs, where the choice of subspace is less restrictive. While motion of a single molecule at a time (dashed yellow lines) shows a small barrier towards dissociation, motion of two molecules at the same time (dashed diagonal green line) results in a high potential energy barrier, making motion of one molecule at a time the most probable scenario. We can see in Fig. 4.2b how this behavior is general for simultaneous motion of two, three, or more molecules. Here, we diagonalize again Eq. (4.7) in two different groups: one of n molecules simultaneously moving (i.e., they all have the same configuration at all times), and other with $N - n$ molecules in the equilibrium position q_{eq}. We see that the energy barrier for simultaneous motion quickly increases with the number of comoving molecules.

This phenomena can be understood easily: there is only a single excitation in the system, which has to be "shared" among N molecules and one cavity mode, i.e., it is coherently distributed over $N + 1$ different states. In the uncoupled system, motion for all but one state proceeds along a ground-state-like surface, introducing a barrier for deviations from the equilibrium position. This feature is necessarily reflected also in the PoPES, where motion of several molecules at a time is strongly suppressed. This phenomena will arise in most typical molecules with locally stable ground states, where motion on the lowest PoPES after photon excitation will proceed mostly along the nuclear coordinate of a single molecule, significantly simplifying the analysis.

We can thus turn our discussion to the full excited-state spectrum, with motion restricted to a single molecule. The uncoupled excited-state PES then consist of N surfaces that follow the ground-state PES along q_1 (the photonically excited PES and $N - 1$ surfaces where a molecule at the equilibrium position is excited), as well as one surface where the moving molecule is excited and the PES thus follows $V_e(q_1)$. In Fig. 4.2c, d, the resulting PoPESs are shown for $N = 5$ and $N = 50$ molecules, respectively, while keeping the Rabi frequency fixed. Note that here the Rabi splitting is defined following Eq. (3.14) for the case of aligned molecules considered here. Therefore, changing the number of molecules while fixing Ω_R corresponds to a change of the effective mode volume (i.e., the single-photon electric field $E_{1\mathrm{ph}}$) for a

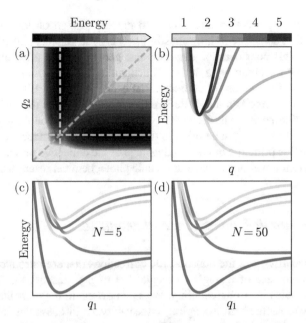

Fig. 4.2 Collective energy landscape in the single-excitation subspace. **a** Lowest excited two-dimension PoPES for motion of two molecules of of $N = 50$ molecules. **b** Lowest excited PoPES for correlatesd simultaneous motion of $n = 1, \ldots, 5$ molecules (for the $N - n$ reamaining molecules in the equilibrium position, with $N = 50$). **c, d** Full PoPES for motion of one molecule for the case of **c** $N = 5$ molecules and **d** $N = 50$ molecules, with identical color code to those in Fig. 4.1. Reprinted with permission from [3]. Copyright 2018 American Chemical Society

constant molecular density. The curves in Fig. 4.2c, d follow the same color code as in Fig. 4.1, indicating the photon fraction and thus the excitonic/polaritonic/photonic nature of each PoPES. The polaritonic parts (in light gray) approximately follow the shape of the ground-state PES. This provides the system some kind of *collective protection* effect, in which after photoexcitation the system now presents a more ground-state-like energy landscape due to the collective coupling of all molecules. This effect can be understood as a generalization to arbitrary PES of the so-called "polaron decoupling" found in Holstein–Tavis–Cummings models [13–15], where the nuclear DoF are treated as pure harmonic oscillators. This effect is analogous to phenomena in J- and H-aggregates, where an excitation is distributed over many molecules not due to coupling with a confined light mode but due to direct inter-molecular interactions [16]. As in molecular aggregates, the similarity between the ground-state PES and the lowest excited PoPES also implies that optical transitions lineshapes should be significantly narrower compared to a bare molecule, due to the fact that the Franck–Condon factors become approximately diagonal (see Sect. 2.2) if the excited PoPES is ground-state-like in a large enough region around the equilibrium position.

The collective protection effect leads to nuclear motion on the PoPES of the single excitation subspace to have mostly the energy dependence of the molecular ground state. This consideration suggest some general design principles in polaritonic chemistry for obtaining a desired functionality by appropriately tailoring the cavity-molecule interaction. In particular, the excited-state PoPES can be obtained by "cutting and pasting" ground-state-like (polaritonic) parts of the surface together with exciton-like parts, with the details determined by the coupling strength and photonic mode frequency in addition to the bare-molecule structure. In Chap. 5 we will illustrate this principle of taking advantage of the collective protection effect to achieve different novel phenomena in organic photochemical reactions.

4.3.2   Polaritonic Nonadiabatic Phenomena

In Sect. 3.2 we discussed the nonadiabatic corrections that are introduced in strong coupling for the case of a single molecule. Let us now discuss this for the case of many molecules in strong coupling. We again disregard the nonadiabatic effects present in the molecule before coupling to the cavity. As discussed in Sect. 2.2, these effects become important when two PESs get close in energy or even degenerate at certain nuclear configurations, which leads to an increase or divergence of the nonadiabatic coupling vector. In the following we study the instances when this take place: avoided crossings and conical intersections.

4.3.2.1   Avoided Crossings

In the full energy landscape of Fig. 4.2c, d we can see an avoided crossing between a purely excitonic PES and a polaritonic one for slightly smaller or larger values of the equilibrium position $q_1 = q_{eq}$. This avoided crossing is clearly seen to become much sharper as the number of molecules is increased. By looking at the lowest excited PoPES, the crossing at $q_1 = q_{cross}$ occurs due to the excited molecule moving sufficiently to fall out of resonance with the photonic mode and thus starts following the uncoupled single molecule PES (exciton-like orange line). Therefore, in the case of $q_1 > q_{cross}$ the PoPES in light gray correspond to coupling between the photonic mode and $N - 1$ molecules, while in $q_1 = q_{eq}$ the polariton is formed with all N molecules. Using a diabatic basis based on these ingredients, i.e., the N polaritonic surfaces that arise from the $N - 1$ ground-state molecules plus the photon and the moving excited molecule, reveals that the effective coupling between these PoPES becomes proportional to the single molecule–photon coupling, which scales as $\sim N^{-1/2}$ for our scenario of the collective Rabi frequency being independent of N. This reduction in coupling can be understood as the distributed excitation having to collapse onto a single molecule as we go over q_{cross}, or equivalently by interpreting the polariton involving the $N - 1$ other molecules as simply a shifted photonic mode coupling to the single-molecule exciton at this position. It should be noted that this reduction is a direct consequence of the collective protection in the system and that further contributes to the molecular stabilization to purely polaritonic states. In our current picture, the uncoupled molecule would immediately tend to dissociate after photoabsorption. However, in strong coupling with a very large collection of molecules the coupling between the excited purely polaritonic diabatic PES and the diabatic uncoupled molecular PES goes to zero as $N \to \infty$, thus making the transition between diabatic PES highly unlikely. In terms of adiabatic surfaces, the nonadiabatic coupling between the two lowest excited PES at $q_1 = q_{\text{cross}}$ is so great that an excited wavepacket will very efficiently transfer between the two so that population does not reach the dissociating region of the PES.

4.3.2.2 Dark-State Collective Conical Intersections

Let us now focus in more detail in the mostly excitonic regions of the energy land-scape close to the bare-molecule excitation energy at equilibrium (see black box in inset of Fig. 4.3). When all molecules are in the same configuration there are $N - 1$ so-called "dark states" (see TC model discussion in Sect. 2.3). In a picture of potential energy surfaces, this means that exactly at this configuration $N - 1$ PoPESs become degenerate. Motion of any degree of freedom (i.e., nuclear motion of any molecule) lift these degeneracies. This can be seen in Fig. 4.3, which corresponds to a zoom of the crossing dark state surfaces for motion of two different molecules. Here, the two sloped surfaces (green and red) roughly correspond to motion of the uncoupled single-molecule excited PES of each of the two molecules, while the orange hori-zontal surface actually corresponds to the remaining $N - 3$ dark PESs, which are completely degenerate for all q_1, q_2 (remember that here we focus on the subspace of restricted motion for $N - 2$ molecules). Along the seams where this surface inter-sect the sloped surfaces, $N - 2$ surfaces are degenerate. These seams (continuous black lines) correspond to either molecule at $q = q_{\text{eq}}$ of collective resonance. The structure discussed here thus gives rise to a high-dimensional hierarchy of hyperdi-mensional surfaces where between 2 and $N - 1$ PoPESs become degenerate, i.e., conical intersection seams of different dimensionality [17–20].

Here it should be noted that these intersections do not correspond simply to intersections of completely decoupled surfaces, but that they actually show nonzero coupling away from the point of intersection (see along the diagonal $q_1 = q_2$, where outside q_{eq} the small coupling lifts the degeneracy between the red and green sur-faces). This interaction is due to the cavity, as the approximate dark states are not completely dark anymore if the perfect degeneracy between emitters is lifted. At the same time, the very small coupling to the cavity implies that the resultant electronic–photonic states are almost purely excitonic and thus their intrinsic linewidth is essen-tially equal to the one of the bare molecule.

One particularly interesting detail here is that these are *collective* conical inter-sections, as they describe nuclear motion of different, possibly spatially separated molecules. This further validates the concept of a "supermolecule" formed from all molecules through collective strong coupling, being this new kind of conical inter-

Fig. 4.3 Zoom on the collective light-induced conical intersection between dark-state PoPESs under motion of two molecules. Reprinted with permission from [3]. Copyright 2018 American Chemical Society

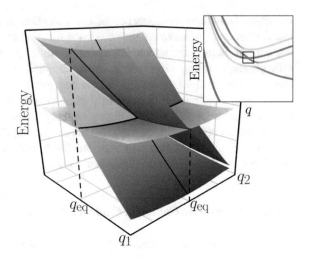

section another example of nontrivial collective effects. Note that the occurrence of this structure is quite robust against inhomogeneous broadening, i.e., shifts of the transition energies of the different molecules. This just leads to slight shifts of the nuclear positions where the different molecular PESs become degenerate and form the conical intersections, but does not destroy their topological properties. In a recent analysis of the dynamics through collective conical intersections [21] it has been demonstrated that these have an important role in the non-radiative energy relaxation rates upper and lower PoPESs, being this governed by the number of coupled molecules [22–24].

4.4 Conclusions

In this chapter we presented an overview of the theory of polaritonic chemistry under strong light–matter coupling based on the picture of PoPES, which generalize the concept of PES to hybrid electron–photon surfaces with a parametric nuclear dependence. In the case of collective strong coupling, more common in experiments, we show the general properties of "collective protection", which lead to PoPESs that can be understood from a principle of "cut and paste" operations combining ground-state-like with exciton-like surfaces. This implies a wide range of freedom for the design of customized PESs that can describe the desired processes. This allows the transformation of the surfaces that govern different photophysical and photochemical phenomena in order to manipulate the resulting product. We here illustrated these concepts using a common energy landscape modeling a chemical scenario of bond dissociation, for the case of coupling to a cavity mode of a single molecule and of a collection of molecules. In the simplest single-molecule scenario we discussed the underlying theory and the consequences of light–matter hybridization in molecules.

In the more complex case of collective strong coupling we present a method to study collective effects with a large number of molecules. Furthermore, we demonstrated a high-dimensional nested structure of collective conical intersection with varying amount of degeneracy induced in the dark states of the system. In the following chapter we present a couple of examples more in order to show the wide variety of possibilities for manipulation of photochemistry that the PoPESs offer.

The PoPES picture opens the possibility of a straightforward interface to existing quantum chemistry methods. With this approach we can make physical predictions of how the energy landscape is altered in realistic strong coupling systems by previously calculating the molecular information with any current available quantum chemistry packages such as Therachem [25, 26]. The bare-molecule structure can then be calculated at each nuclear configuration for each molecule separately. Then, it is possible to treat the light–matter coupling within a small Hamiltonian involving only a few relevant states per molecule, similar to that in existing excitonic models [27, 28]. For the PoPES approach, this effective decoupling between the "chemical" and "quantum optical" parts of the calculation allows the use of well-known approaches such as QM/MM (quantum mechanics/molecular mechanics) for treating big molecular systems. Here, nuclear motion on the PoPESs is treated classically, with nonadiabatic couplings introduced through surface hopping algorithms. The clear parallelizability of this approach has allowed the treatment of up to 1600 rhodamine molecules (within the single-excitation subspace) and their surrounding solvent, corresponding to 43 200 QM and 17 700 800 MM atoms in total [11].

References

1. Buhmann SY (2007) Casimir-Polder forces on atoms in the presence of magnetoelectric bodies. Thesis (PhD), Friedrich-Schiller-Universität Jena
2. Spano FC (2015) Optical microcavities enhance the exciton coherence length and eliminate vibronic coupling in J-aggregates. J Chem Phys 142
3. Feist J, Galego J, Garcia-Vidal FJ (2018) Polaritonic chemistry with organic molecules. ACS Photonics 5:205
4. Lidzey DG, Bradley DDC, Armitage A, Walker S, Skolnick MS (2000) Photon-mediated hybridization of Frenkel excitons in organic semiconductor microcavities. Science 288:1620
5. Rodriguez SRK, Feist J, Verschuuren MA, Garcia Vidal FJ, Gómez Rivas J (2013) Thermalization and cooling of plasmon-exciton polaritons: towards quantum condensation. Phys Rev Lett 111
6. Zengin G, Wersäll M, Nilsson S, Antosiewicz TJ, Käll M, Shegai T (2015) Realizing strong light-matter interactions between single-nanoparticle plasmons and molecular excitons at ambient conditions. Phys Rev Lett 114:157401
7. Chikkaraddy R, de Nijs B, Benz F, Barrow SJ, Scherman OA, Rosta E, Demetriadou A, Fox P, Hess O, Baumberg JJ (2016) Single-molecule strong coupling at room temperature in plasmonic nanocavities. Nature 535:127
8. Schwartz T, Hutchison JA, Léonard J, Genet C, Haacke S, Ebbesen TW (2013) Polariton dynamics under strong light-molecule coupling. ChemPhysChem 14:125
9. Velha P, Picard E, Charvolin T, Hadji E, Rodier J, Lalanne P, Peyrade D (2007) Ultra-high Q/V Fabry-Perot microcavity on SOI substrate. Opt Express 15:16090

10. Akahane Y, Asano T, Song B-S, Noda S (2003) High-Q photonic nanocavity in a two-dimensional photonic crystal. Nature 425:944
11. Luk HL, Feist J, Toppari JJ, Groenhof G (2017) Multiscale molecular dynamics simulations of polaritonic chemistry. J Chem Theory Comput 13:4324
12. Garraway BM (2011) The Dicke model in quantum optics: Dicke model revisited. Philos Trans R Soc A Math Phys Eng Sci 369:1137
13. Herrera F, Spano FC (2016) Cavity-controlled chemistry in molecular ensembles. Phys Rev Lett 116:238301
14. Herrera F, Spano FC (2018) Theory of nanoscale organic cavities: the essential role of vibration-photon dressed states. ACS Photonics 5:65
15. Wu N, Feist J, Garcia-Vidal FJ (2016) When polarons meet polaritons: exciton-vibration interactions in organic molecules strongly coupled to confined light fields. Phys Rev B 94:195409
16. Kasha M (1963) Energy transfer mechanisms and the molecular exciton model for molecular aggregates. Radiat Res 20:55
17. Worth GA, Cederbaum LS (2004) Beyond Born-Oppenheimer: molecular dynamics through a conical intersection. Annu Rev Phys Chem 55:127
18. Levine BG, Martínez TJ (2007) Isomerization through conical intersections. Annu Rev Phys Chem 58:613
19. Domcke W, Yarkony DR, Köppel H (eds) (2004) Conical intersections: electronic structure, dynamics and spectroscopy. Advanced series in physical chemistry, vol 15. World Scientific Publishing Co. Pte. Ltd
20. Domcke W, Yarkony DR (2012) Role of conical intersections in molecular spectroscopy and photoinduced chemical dynamics. Annu Rev Phys Chem 63:325
21. Vendrell O (2018) Collective Jahn-Teller interactions through light-matter coupling in a cavity. Phys Rev Lett 121:253001
22. Pino JD, Feist J, Garcia-Vidal FJ (2015) Quantum theory of collective strong coupling of molecular vibrations with a microcavity mode. New J Phys 17:53040
23. Sáez-Blázquez R, Feist J, Fernández-Domínguez A, García-Vidal F (2018) Organic polaritons enable local vibrations to drive long-range energy transfer. Phys Rev B 97:241407
24. Ribeiro RF, Martínez-Martínez LA, Du M, Campos-Gonzalez-Angulo J, Yuen-Zhou J (2018) Polariton chemistry: controlling molecular dynamics with optical cavities. Chem Sci 9:6325
25. Ufimtsev IS, Martinez TJ (2009) Quantum chemistry on graphical processing units. 3. Analytical energy gradients, geometry optimization, and first principles molecular dynamics. J Chem Theory Comput 5:2619
26. Titov AV, Ufimtsev IS, Luehr N, Martinez TJ (2013) Generating efficient quantum chemistry codes for novel architectures. J Chem Theory Comput 9:213
27. Sisto A, Glowacki DR, Martinez TJ (2014) Ab initio nonadiabatic dynamics of multichromophore complexes: a scalable graphical-processing-unit-accelerated exciton framework. Acc Chem Res 47:2857
28. Sisto A, Stross C, van der Kamp MW, O'Connor M, McIntosh-Smith S, Johnson GT, Hohenstein EG, Manby FR, Glowacki DR, Martinez TJ (2017) Atomistic non-adiabatic dynamics of the LH2 complex with a GPU-accelerated ab initio exciton model. Phys Chem Chem Phys 19:14924

Chapter 5
Manipulating Photochemistry

5.1 Introduction

The possibility of manipulating photochemical reactions exploiting strong light–matter coupling, as demonstrated in several experiments [1–3], holds great interest for many different fields of science. In this chapter we apply the theory developed in previous chapters in order to understand and predict different chemical changes in organic photochemistry. We treat two different molecular models that represent typical simple photoisomerization reactions. In Sect. 5.2 we discuss the possibility of taking advantage of the collective protection effect introduced in Sect. 4.3 in order to suppress photoisomerization reactions. We study the single-molecule dynamics in strong coupling and the decrease of the reaction rate with the number of molecules. We then investigate the system with two excitations and discuss its effect on the reaction suppression and the connection to polariton–polariton interactions. Then, in Sect. 5.3 we demonstrate the possibility of engineering the PoPES to achieve novel reaction pathways by properly tuning the system in strong coupling. We show how it is possible to increase the quantum yield of a photochemical reaction, and even overcome the second law of photochemistry in the case of many molecules.

5.2 Suppressing Photochemical Reactions

An organic molecule can undergo an structural change after absorption of a photon. This process is known as photoisomerization, a mechanism of great importance in many biological systems such as the human eye [4]. It presents plenty of possible technological applications in solar energy storage [5] and as optical switches, memories, and actuators [6, 7]. However, it can also have detrimental effects, such as limiting the efficiency of organic solar cells [8] or opening important damage pathways in DNA under solar radiation [9, 10]. While sometimes these effects can be

© The Editor(s) (if applicable) and The Author(s), under exclusive license
to Springer Nature Switzerland AG 2020
J. Galego Pascual, *Polaritonic Chemistry*, Springer Theses,
https://doi.org/10.1007/978-3-030-48698-3_5

avoided by shielding the system from light, this is not a viable pathway when the system precisely relies on the interaction with external light, such as in the case of solar cells.

In this section we will show that it is possible to suppress photoisomerization by strongly coupling the relevant molecules to confined light modes. The photochemical process is governed by the excited state PES, which is modified under strong coupling, thus influencing the dynamics that lead to structural changes. The PoPES that controls the reaction develops a new minima in which the excited wavepackets are trapped after photoabsorption. We observe the dynamics in the single-molecule scenario, and then we discuss how we can exploit the collective protection effect presented in Sect. 4.3 in order to enhance the suppression of the reaction.

We treat a general molecular model that can represent a variety of commonly studied photoisomerization reactions, such as *cis-trans* isomerization of stilbene, azobenzene or rhodopsin [4, 11, 12] (corresponding to rotation around a C=C or N=N double bond, as sketched in Fig. 5.1a), or ring-opening and ring-closing reactions in diarylethenes [6]. The model molecule describes nuclear motion on ground and excited electronic PES along a single reaction coordinate q.

The adiabatic electronic PESs of the bare molecule are constructed in terms of diabatic surfaces $V_A(q)$ and $V_B(q)$, shown in Fig. 5.1b, which are coupled to each

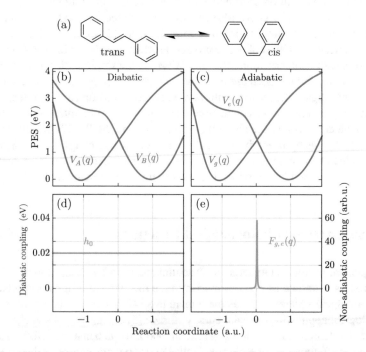

Fig. 5.1 a Sketch of the *cis-trans* isomerization reaction of stilbene. **b** Diabatic and **c** adiabatic PES for the model molecule. **d** Constant coupling between the diabatic surfaces. **e** Nonadiabatic coupling at $q \approx 0$ between adiabatic surfaces $V_g(q)$ and $V_e(q)$

other with a small coupling $h_0 = 0.02$ eV assumed constant in space. This gives the following electronic Hamiltonian:

$$\hat{H}_e(q) = \begin{pmatrix} V_A(q) & h_0 \\ h_0 & V_B(q) \end{pmatrix}. \tag{5.1}$$

Diagonalization of $\hat{H}_{el}(q)$ returns the ground and excited state PES of Fig. 5.1c, $V_g(q)$ and $V_e(q)$, together with the adiabatic electronic wavefunctions. This also gives access to the nonadiabatic coupling that controls the transition between ground and excited surfaces at $q \approx 0$ (see Fig. 5.1e), given by $F_{i,j}(q) = \langle i(q)| \partial_q |j(q)\rangle$, where $i, j \in \{e, g\}$ and $|i(q)\rangle$ represent the adiabatic electronic states. We note that nonadiabatic transitions in "real" molecules typically involve conical intersections (see the corresponding discussion in Sect. 2.2), which only occur in multi-dimensional systems; however, the details of this transition do not influence the results presented.

We take into account only one DoF, while all the others are assumed to be fully relaxed such that the excited PES represents the minimum energy path that governs the reaction (see Sect. 2.2). The ground state PES, $V_g(q)$ (blue line), possesses minima at $q = q_0 \approx -1.05$ a.u. and $q \approx 1$ a.u., corresponding to the stable (e.g., trans-) and metastable (e.g., cis-) isomers, respectively. They are separated by a barrier with a maximum at $q \approx 0$ accompanied by an avoided crossing between the ground and excited state PES, $V_e(q)$ (orange line). In order to ensure a large quantum yield for photoisomerization in the bare molecule, we choose a very narrow avoided crossing (with energy splitting 39 meV, smaller than the width of the lines in Fig. 5.1c).

Besides the bare-molecule PESs, we need to set the molecular dipole moment operator $\hat{\mu}$, which determines the coupling to the quantized light mode and the absorption of the system. For simplicity we set $\hat{\mu}$ to be purely offdiagonal in the adiabatic basis, i.e., $\mu_{gg} = \mu_{ee} = 0$. The ground-excited transition dipole moment μ_{eg} is approximately constant close to the stable geometries, but changes rapidly close to the nonadiabatic transition due to the sudden polarization effect [13]. We thus choose $|\mu_{eg}(q)| \propto \arctan(q/q_m)$, with $q_m = 0.625$ representing the length scale on which $\mu_{eg}(q)$ changes. As discussed below, the specific shape of $\mu_{eg}(q)$ does not strongly affect the results presented here.

Therefore the complete molecular Hamiltonian is then given by

$$\hat{H}_{mol}(q) = \frac{\hat{P}^2}{2M_q} + \hat{V}(q) + \hat{\Lambda}(q), \tag{5.2}$$

where $\hat{P}$ is the (diagonal) nuclear momentum operator, M_q is the effective mass for the nuclear coordinate q, $\hat{V}(q)$ is the (diagonal) potential operator in the adiabatic basis, and $\hat{\Lambda}(q)$ is the operator of offdiagonal (nonadiabatic) couplings as defined in Eq. (2.32).

5.2.1   Single Molecule Dynamics

In order to study the dynamics, we need to calculate the population evolution both in
the uncoupled and in the strongly coupled systems by solving the Schrödinger equa-
tion $i\partial_t |\psi(t)\rangle = \hat{H}_{tot} |\psi(t)\rangle$, where $\hat{H}_{tot}$ is the total Hamiltonian without invoking
the Born–Oppenheimer approximation, i.e., including all nonadiabatic terms, both
in the coupled and uncoupled case. We use a finite-element discrete variable representation [14, 15] for the nuclear coordinate q, as well as a Fock basis for the cavity
photon mode. Note that in the strong coupling regime the nonadiabatic couplings in
the polaritonic basis are given by new terms $\hat{\Lambda}_{SC}$ due to the change of basis, as well
as the original bare-molecule nonadiabatic couplings $\hat{\Lambda}(q)$ transformed to the new
eigenstate basis. The initial wavefunction is given by direct promotion of the ground-
state nuclear wavepacket to the lowest excited state (i.e., $|e\rangle$ for no coupling and $|LP\rangle$
under strong coupling), filtered by the q-dependent transition dipole moment μ_{eg}.
This approximately corresponds to the initial state that would be obtained after excitation by an ultrashort laser pulse tuned to the excited state energy around the nuclear
equilibrium position.

In Fig. 5.2a, d we show the population transfer from the excited to the ground
state in the case of the uncoupled molecule. When the wavepacket encounters the
avoided crossing, it undergoes an efficient nonadiabatic transition (i.e., it follows the
diabatic surfaces, see Fig. 5.1b). In this scenario, the bare model molecule undergoes
rapid photoisomerization, with the nuclear wavepacket reaching the second isomer
($q > 0$) within a few hundred fs.

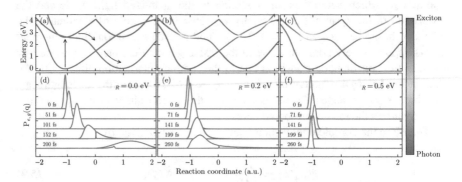

Fig. 5.2 Suppression of photoisomerization under strong coupling for a single molecule. **a–c**
Ground (blue) and excited (purple-orange color scale) potential energy surfaces of the model
molecule coupled to a quantized light mode ($\omega_c = 2.65$ eV), with the light-matter coupling strength
Ω_R increasing from **a** to **c**. The continuous color scale encodes the nature of the hybridized excited
PES. **d–f** Time propagation of the nuclear wavepacket after sudden excitation to the lowest excited
PES (lower polariton for $\Omega_R > 0$), shown separately for the parts in the lower polariton surface
(orange) and the ground state surface (blue) reached through the nonadiabatic transition at $q = 0$.
Contributions in the upper polariton surface are negligible and not shown

In contrast, when the system enters strong coupling, photoisomerization in a single molecule is suppressed. To show this, we rely on the theoretical framework we introduced in Chap. 4. The electron–photon Hamiltonian is given by

$$\hat{H}_{SC} = \omega_c \hat{a}^\dagger \hat{a} + \hat{V}(q) + \hat{\boldsymbol{\mu}}(q) \cdot \mathbf{E}_{1ph}(\hat{a}^\dagger + \hat{a}), \tag{5.3}$$

where ω_c is the quantized light mode frequency, and $\mathbf{E}_{1ph}$ is the electric field amplitude of a single confined photon. Without the light–matter coupling term, the photonically excited surface describes the motion of a ground-state molecule with an (uncoupled) photon present in the cavity, and is thus simply a copy of the molecular ground state shifted upwards by the photon energy, $V_c(q) = V_g(q) + \omega_c$ (purple curve in Fig. 5.2a). When coupling is turned on, PoPES are formed, presenting both photonic and excitonic character (see Fig. 5.2b, c). The splitting between the PoPES around the equilibrium ($q_0 \approx -1.05$ a.u.) is approximately equal to the Rabi frequency $\Omega_R = 2\boldsymbol{\mu}_{eg}(q_0) \cdot \mathbf{E}_{1ph}$.

Importantly, we observe in Fig. 5.2b, c that the lower PoPES develops a deeper and deeper minimum as the coupling is increased. This has two primary reasons: first, the light–matter coupling is most effective when $V_c(q)$ and $V_e(q)$ are close, "pushing down" the lower polariton at the equilibrium position. At regions of larger detuning, the "polaritonic" PES are almost identical to the uncoupled ones. Second, the local shape of the PoPES becomes a mixture of the two uncoupled PES in regions where they hybridize significantly. Since the photonic surface $V_c(q)$ behaves like the ground-state PES, this additionally supports the formation of a local minimum in the PoPES. In combination, this leads to the formation of a reaction barrier against isomerization as the coupling is increased. At intermediate coupling, where no barrier is formed yet, the reaction is slowed down, but not suppressed (see Fig. 5.2b, e). Once the coupling becomes sufficiently large, a barrier appears and the excited wavepacket is trapped in the local minimum, such that isomerization becomes almost completely suppressed (see Fig. 5.2c, f). The initial wavepacket on the lower polariton surface in our calculations is started by a sudden transition and thus includes all vibrationally excited states that are reachable from the ground state through a dipole transition. If the coherent excited wavepacket is successfully trapped without undergoing ultrafast isomerization (as in Fig. 5.2c, f), the ultimate fate of the molecule will be determined by two additional effects: on the one hand, the excited wavepacket will thermalize within the lower PoPES on typical timescales of picoseconds. While the exact values depend on the details of the system, we note that for the model molecule treated here, the barrier height of ≈ 65 meV in Fig. 5.2c is much larger than the thermal energy $k_B T \approx 26$ meV at room temperature, preventing isomerization, according to transition state theory (TST, see Sect. 2.2). On the other hand, the excited-state wavepacket will simultaneously decay both by radiative and nonradiative processes with timescales typically dominated by the photonic part of the PoPES, ranging from tens of femtoseconds for plasmonic resonances to picoseconds and longer for dielectric structures.

Note that while the upper PoPES appears even more stable than the lower one in this model, this is an artifact of the restriction to one degree of freedom, with all other degrees of freedom relaxed to their local minimum. This implies that the lower PoPES indeed corresponds to the lowest-energy excited state, such that the restriction to one coordinate is well-justified. In contrast, the upper polariton surface can possess efficient relaxation pathways to the lower polariton along orthogonal degrees of freedom, and indeed, upper polaritons are known to decay relatively quickly within the excited-state subspace [16, 17].

We have thus shown that strong coupling of a single molecule to a confined light mode can strongly suppress photoisomerization reactions and stabilize the molecule. The experimental realization of single-molecule strong coupling [18] proves that this could indeed be a viable pathway towards manipulation of single molecules. At the same time, most experiments achieving strong coupling with organic molecules have exploited *collective* coupling [19, 20], in which $N \gg 1$ molecules coherently interact with a single mode, leading to an enhancement of the total Rabi frequency by a factor of $\sqrt{N}$ (see Eq. (3.14) and TC model).

As presented in Sect. 4.3, when in collective strong coupling, the system will experience the *collective protection* effect. In this scenario it is not immediately clear whether this will be detrimental or beneficial to the suppression of the reaction. As has recently been shown, many observables corresponding to "internal" degrees of freedom of the molecules are only affected by the single-molecule coupling strength and thus not strongly modified under collective strong coupling [21, 22]. We therefore in the following explicitly check whether this suppression effect is "washed out" by the presence of the other molecules by explicitly studying a collection of N molecules in strong coupling.

5.2.2 Collective Suppression

In order to treat collective strong coupling involving N molecules and a single confined light mode, we again restrict ourselves to the zero- and single-excitation subspace. As we have seen in the theory Sect. 4.3, the molecules now have N total nuclear degrees of freedom, described by the vector $\mathbf{q} = (q_1, \ldots, q_N)$, and the PES accordingly become N-dimensional surfaces. Diagonalization of the full Hamiltonian of Eq. (4.4) gives $N + 1$ polaritonic surfaces, which describe the collective coupled motion of all molecules. In principle, this could induce, e.g., collective transitions in which multiple molecules move in concert, but an explicit analysis of all the nuclear DoF is computationally not feasible. We therefore focus on the subspace of the full Hilbert space where many molecules share the same configuration by using the collective spin operators defined in Eq. (4.6). As in the previous chapter, we do it for motion of two molecules while keeping the remaining $N - 2$ molecules in the equilibrium position ($q_j = q_0$ for $j > 2$), shown in Fig. 5.3a. Again, we compute the energy profile for simultaneous motion of $n = 1, \ldots, 5$ molecules, shown in Fig. 5.3b. We see that the same arguments of Sect. 4.3 are repeated, as the collective protection effect makes that motion of more than one molecule takes place in ground-state potential wells, i.e., along steep potential barriers.

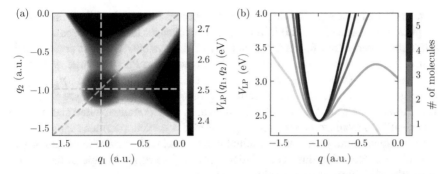

Fig. 5.3 Many-molecule potential energy surfaces under strong coupling. **a** Lower polariton PES for $N = 50$ molecules, under motion of molecules 1 and 2, with all others held in the equilibrium position q_0. **b** Energy reaction path corresponding to simultaneous motion of several molecules. In both panels, the photonic mode frequency is $\omega_c = 2.65$ eV, while the (collective) Rabi frequency is fixed to $\Omega_C = \sqrt{N}\Omega_R = 0.5$ eV

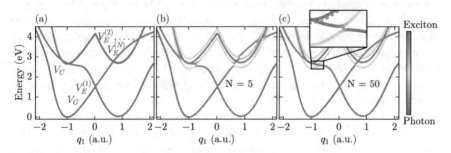

Fig. 5.4 Many-molecule potential energy surfaces under strong coupling. **a–c** All potential energy surfaces under motion of only molecule 1, for no light-matter coupling (**a**), and under strong coupling for $N = 5$ (**b**) and $N = 50$ (**c**) molecules. In all panels, the photonic mode frequency is $\omega_c = 2.65$ eV, while the (collective) Rabi frequency is fixed to $\Omega_C = \sqrt{N}\Omega_R = 0.5$ eV

We thus analyze the coupled states under motion of only the first molecule q_1, fixing all other molecules to the ground-state equilibrium position ($q_j = q_0$ for $j > 1$). The corresponding surfaces are shown in Fig. 5.4. When the light–matter coupling is zero (Fig. 5.4a), the surface $V_E^{(1)}(\mathbf{q})$ behaves like $V_e(q_1)$, while all other surfaces (corresponding to photonic excitation, or excitation of a "stationary" molecule $j > 1$) appear like copies of the ground-state PES $V_g(q_1)$ shifted in energy. The PoPES for varying numbers of molecules are shown in Fig. 5.3b, c. We keep the total Rabi frequency constant (corresponding to a scaling of the single-photon field strength with $N^{-1/2}$). Close to equilibrium ($q_1 \approx q_0$), the $N + 1$ surfaces can be clearly classified into a lower and upper polaritonic PES (light gray), which show significant hybridization with the photonic mode, as well as $N - 1$ "dark" surfaces (orange) that are almost purely excitonic.

As the number of molecules is increased the collective protection effect acquires more relevance, and thus the local minimum of the lower PoPES (the lowest light gray surface) close to the equilibrium position $\mathbf{q}_0$ becomes more and more reminiscent of the pure ground-state PES. This introduces a set of changes to the landscape of excited states that utterly transforms the dynamics of the system. The first consequence is that the similarity of the ground and lower polariton PES for large N implies that the Franck–Condon factors become approximately diagonal (i.e., $\boldsymbol{\mu}_{\mathrm{LP},\mathrm{g}}(\mathbf{q}) \approx \boldsymbol{\mu}_{\mathrm{LP},\mathrm{g}}(\mathbf{q}_0)$ over the width of the vibrational ground-state wavepacket, see Sect. 4.3). Thus, transitions from the overall ground state to vibrationally excited states in the lower PoPES become more and more suppressed. Photoexcitation then cannot change the vibrational state, such that the excited wavepacket will be close to its vibrational ground state. Note that this occurs independent of the general shape of the single-molecule offdiagonal transition dipole moment.

Additionaly, this likeness to the ground-state PES will also generate an energy barrier to photoisomerization increasingly higher with the number of molecules. For the cases shown in Fig. 5.4, the barrier height reaches ≈ 117 meV for $N = 5$ and ≈ 156 meV for $N = 50$ molecules, well above the thermal energy at room temperature. Therefore, after photoexcitation, a ground-state wavepacket can then thermalize (on typical time scales of picoseconds at room temperature), with the lifetime for passing over the barrier determined by the probability of gaining enough energy from the bath to overcome the barrier. We can now obtain an estimate of the lifetime based on TST. This estimate should be taken with some caution, as there are at least two features of the polaritonic system considered here that differ from the situation treated by standard TST.

First of all, there is not just a single energetic barrier that has to be overcome, but one for motion of every molecule with all $N - 1$ others close to equilibrium. This enhances the probability of overcoming the barrier by a factor of N. However, in the diabatic picture, the number of molecules also introduces a scaling of $1/N$ to the transition probability at the avoided crossing (see isent in Fig. 5.4c) to the excitonic PES that leads to photoisomerization. Therefore the two effects discussed approximately cancel each other. We thus assume that TST provides a useful estimate of the excited-state lifetime.[1] Under the condition that the photon frequency is fixed to stay close to resonance at the equilibrium position, the barrier height in the lowest PoPES depends on two parameters: the Rabi frequency and the number of molecules. Their combined effect on the energy barrier and the corresponding lifetime is shown in Fig. 5.5, which demonstrates that increasing N leads to higher barriers, with the value saturating for a given Rabi splitting. Alternatively, larger Rabi frequencies and the associated reduction in the minimum energy of the lowest PoPES lead to effectively higher barriers and thus a more efficient suppression of the photoisomerization reaction. The associated lifetimes range from about one picosecond to about 10 ns depending on parameters. The final fate of an excited wavepacket will thus depend on the competition between two time scales: that of the vibrational wavepacket trapped

---

[1] It should be noted that the lifetimes calculated here are an estimate, and a more precise calculation would require to compute the formally exact quantum reaction rates as discussed in Sect. 2.2 in order to properly include effects such as tunneling or recrossings. This nevertheless does not strongly affect the general suppression result.

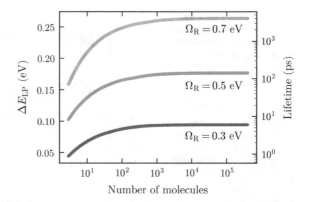

Fig. 5.5 Energy barriers versus number of molecules for different values of Rabi splitting. The right axis shows the equivalen lifetime predicted through transition-state theory. Adapted with permission from [23]. Copyright 2018 American Chemical Society

inside the local potential well in the lower PoPES, as well as that of the polaritonic state against relaxation, which is typically dominated by the photonic fraction of the polariton.

5.2.3 Beyond the Single-Excitation Subspace

Up to now, we have only discussed the PoPES within the zero- and single-excitation subspace. While this is the subspace probed under weak excitation (linear response) in experiment, the nonlinear response of polaritonic systems is a topic of great current interest. It becomes relevant in, among others, transient absorption measurements [1, 24], nonlinear optics setups [25–27], and studies of polariton lasing and condensation [28–32]. We here focus on the two-excitation subspace and investigate whether we still observe the collective protection effect and the resulting suppression of the phosotoisomerization, and whether we observe any effective polariton–polariton interactions leading to correlated motion. As in the rest of the section, we neglect direct dipole–dipole interactions between the molecules, such that saturation is the only source of nonlinearities or effective polariton–polariton interactions (as typically observed in organic-based polaritonic systems due to the localized nature of Frenkel excitons [28]). When the number of molecules is much larger than the number of excitations, it is expected that the system bosonizes, i.e., that the response becomes linear and polaritons become approximately independent of each other [33]. However, nonlinearities can survive even for surprisingly large values of N under strong coupling conditions [34]. In this case, the use of collective spin operators becomes essential for its computational treatment, as it keeps the problem easily tractable even for relatively large numbers of molecules (where a naive approach would scale with N^2).

We now calculate the PoPES for up to two excitations with $N = 50$, shown in Fig. 5.6 for the region $q_1 \lesssim 0$. The large number of surfaces seen in the two-excitation subspace can be approximately qualified within an independent-particle

Fig. 5.6 PoPES up to the two-excitation subspace for the treated model molecule, for $N = 50$ molecules, photonic mode frequency $\omega_c = 2.65$ eV, and (collective) Rabi frequency fixed to $\Omega_C = 0.5$ eV. The two color scales indicate the photon fraction in the single- and two-excitation subspace, respectively. Reprinted with permission from [23]. Copyright 2018 American Chemical Society

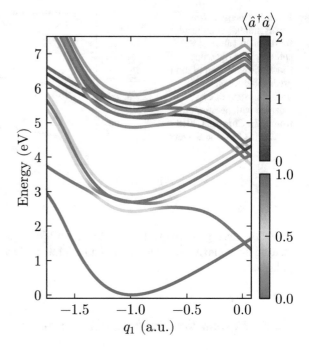

picture (expected to be exact in the limit $N \to \infty$), corresponding to, for example, excitation of 2 lower polaritons, or one lower and one upper polariton. The color scale within the two-excitation subspace again measures the photonic contribution to each state, now spanning from $\langle n_{ph} \rangle = 0$ (two excitons, dark orange) through dark gray (one exciton, one photon) to $\langle n_{ph} \rangle = 2$ (two photons, dark purple).

In the following, we will focus on the lowest PoPES within the two-excitation manifold, which we label as $V_{2LP}(q_1, q_2, \ldots, q_N)$ as it corresponds approximately to two lower polaritons close to equilibrium. Note that for the two-state molecules considered here, isomerization after double excitation (i.e., after absorption of two photons) in the uncoupled system corresponds to two independent excitons on separate molecules, with motion proceeding on the surface $V_{2e}(q_1, q_2) = V_e(q_1) + V_e(q_2)$. This implies that, in contrast to the single-excitation subspace, concerted motion of two molecules (e.g., along $q_1 = q_2$) is not a priori suppressed under strong coupling. In the limit $N \to \infty$, the lowest surface should again support independent motion, but now on polaritonic surfaces. Consequently, a cut where only two molecules move should approximately fulfill $V_{2LP}(q_1, q_2, q_0, \ldots, q_0) \approx V_{LP}(q_1) + V_{LP}(q_2)$.

This is studied in Fig. 5.7, which shows two cuts through V_{2LP}, one in which only q_1 is varied (red solid line) and one in which $q_1 = q_2$ are varied together (green solid line). In addition, it shows the independent-particle limit of $2V_{LP}(q_1)$ (dashed yellow line). In all three cases, all remaining molecules are fixed to the equilibrium position. The plots are restricted to the region of interest $q \lesssim -0.5$ a.u., with subplots showing the cases $N = 5$ (a) and $N = 100$ (b). For the case of $N = 5$ molecules,

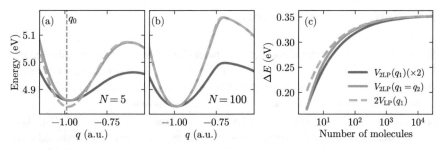

Fig. 5.7 a, b Comparison of the PoPES in the double-excitation subspace for motion of one molecule (red line) and two molecules (green line), and twice the lower PoPES in the single-excitation subspace for motion of one molecule (orange dashed line), for $N = 5$ and $N = 100$ coupled molecules, respectively. **c** Dependence of the energy barrier for double photoisomerization with the number of molecules for the reaction paths corresponding to the PoPES in (**a, b**). Reprinted with permission from [23]. Copyright 2018 American Chemical Society

it is clearly visible that simultaneous motion of two molecules has a slightly lower barrier than would be expected in the independent-particle limit due to a noticeable blueshift around the equilibrium position, indicating effective polariton–polariton interactions. These differences disappear for large enough N and are barely visible for the case $N = 100$ shown in Fig. 5.7b.

In order to clearly distinguish whether simultaneous motion of several molecules is favored compared to the independent-particle limit, we now directly compare the barrier heights for two-molecule reactions for the different cases studied here in Fig. 5.7c. Each line corresponds to the same case in Fig. 5.7a, b, with the difference that the energy barrier ΔE for motion of one molecule in the two-excitation subspace has been multiplied by 2 for the sake of comparison. As expected, all correlations disappear for large numbers of molecules, with the barrier heights converging to the same value. However, even for a considerable number of molecules such as $N = 100$ the correlations are non-negligible (remember that transition rates approximately depend exponentially on barrier height, according to TST), suggesting that polaritonic chemistry could possess subtle non-bosonic response even for mesoscopic numbers of molecules, similarly as recently found for photon correlations [34, 35]. In particular, for the model studied here, the barrier for simultaneous motion of two molecules after double excitation of the system $V_{2LP}(q_1, q_2)$ is slightly smaller than expected from an independent-particle model ($2V_{LP}(q_1)$). Interestingly, motion of just a single molecule in the two-excitation subspace is even less suppressed, with the barrier consistently less than twice as high. It should be noted that the subtle effects found for the specific model discussed here will of course be challenging to measure experimentally, but they could point a way towards more pronounced polariton–polariton interaction effects in polaritonic chemistry.

5.3   Enhancing Photochemistry

We now focus on one particular characteristic of photochemistry: the so-called Stark–Einstein law, which states that "one quantum of light is absorbed per molecule of absorbing and reacting substance" [36]. This means that the quantum yield $\phi = \frac{N_{prod}}{N_{phot}}$ of the reaction, which describes the percentage of molecules that end up in the desired reaction product per absorbed photon, has a maximum value of 1. This limit can be overcome in some specific cases, such as in photochemically induced chain reactions [37–39], or in systems that support singlet fission to create multiple triplet excitons (and thus electron-hole pairs) in solar cells [40, 41].

In this section we demonstrate a novel and efficient approach to circumvent the second law of photochemistry, based on exploiting the collective nature of the new eigenstates of a collection of molecules strongly coupled to confined light modes. This can allow many molecules to undergo a photochemical reaction after excitation with just a single photon, thus achieving an effective quantum yield larger than 1. This obviously cannot lead to a violation of conservation of energy, therefore we investigate a class of exothermic reactions that release energy, i.e., where the initial state before photoabsorption has higher energy than the final state after the reaction has concluded. We focus on a class of model molecules with a structure as proposed for use in solar energy storage [5, 42, 43], again described within a simplified model treating a single reaction coordinate, as shown in Fig. 5.8.

In the model molecule, the PES associated with the electronic ground state contains two local minima: a stable ground-state configuration (at $q = q_s \approx 0.8$ a.u.) and a metastable configuration (at $q = q_{ms} \approx -0.7$ a.u.) that contains a stored energy of about 1 eV. The activation barrier for thermal relaxation from the metastable configuration to the global minimum has a height of more than 1 eV, leading to a lifetime on the order of days or even years for the metastable configuration according to transition state theory, and thus making it interesting for solar energy storage. In addition,

Fig. 5.8 Potential energy surfaces of a molecule presenting a general photoisomerization reaction where one isomer has higher energy than the other, both of them with roughly equal reaction quantum yield after absorption of a photon. Adapted with permission from [44]. Copyright 2017 American Physical Society

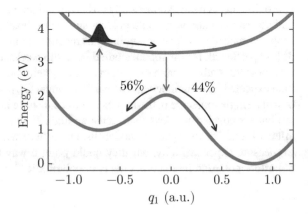

Table 5.1 Molecular model parameters

	E_i (eV)	A_i (eV/a_0^2)	q_i (a_0)	h_i (eV)
v_g	1.34	5.0	−0.75	2.0
w_g	0.29	5.0	0.85	
v_e	11.65	5.0	−1.15	14.0
w_e	10.15	4.0	1.25	

the molecule possesses an excited state PES with a relatively flat minimum close to the ground transition state.

Let us first describe the molecular model in more detail. The two adiabatic PES (ground state and first excited state) of the single bare molecule are both constructed independently from two coupled harmonic potentials as follows:

$$V_i(q) = \frac{1}{2}\left(v_i(q) + w_i(q) - \sqrt{4h_i^2 + [v_i(q) - w_i(q)]^2}\right), \qquad (5.4)$$

where $i \in \{g, e\}$ indicates either the ground state or excited state PES and with

$$v_i(q) = E_i^{(v)} + A_i^{(v)}\left(q - q_i^{(v)}\right)^2, \qquad (5.5a)$$

$$w_i(q) = E_i^{(w)} + A_i^{(w)}\left(q - q_i^{(w)}\right)^2. \qquad (5.5b)$$

Each of the PES (ground and excited state) is then described by 7 parameters, 3 each for $v_i(q)$ and $w_i(q)$, as well as a coupling h_i. Their values are given in Table 5.1. While there is a relatively large number of free parameters that control the molecular structure, we have checked that the results presented below are insensitive to small variations as long as the general shape of the PES is maintained.

In addition to the PES themselves, the properties of the strongly coupled light–matter system depend on the transition dipole moment $\mu_{eg}(q) = \langle e(q)|\hat{\mu}|g(q)\rangle$ between the ground and excited state, which determines the coupling strength to the photon mode. As discussed in the previous section, the q-dependence of the transition dipole moment is typically relatively smooth close to local minima, but can change rapidly close to regions of strong nonadiabaticity, which are absent in the current model. Furthermore, as we have seen in the previous section, the collective protection effect ensures that the effective transition dipole moment between the ground state and hybridized parts of the excited polaritonic PES becomes almost independent of q for $N \gg 1$ for motion of one molecule. For simplicity, we therefore use a q-independent $\mu_{eg}(q)$ in this section.

We finally discuss the decay of an isolated molecule to the ground state after excitation, which determines the bare-molecule quantum yields. For simplicity, we assume that the excitation decays purely radiatively, implying that nonadiabatic effects in the bare molecule are negligible. The fluorescence quantum yield $\gamma_r/(\gamma_r + \gamma_{nr})$ is then close to 1, where γ_r (γ_{nr}) is the radiative (nonradiative) decay rate from the excited state. Furthermore, since radiative lifetimes are much longer than vibrational relaxation, we can assume that the wavepacket in the excited PES has reached thermal equilibrium before fluorescence. We can thus approximate the associated wavepacket by the vibrational ground state $\chi_e^{(0)}(q)$, obtained by diagonalizing the adiabatic Hamiltonian $\frac{\hat{P}^2}{2M_q} + V_e(q)$, where $\hat{P} = -i\frac{\partial}{\partial q}$ is the nuclear momentum operator and $M_q = 550$ Da is the effective mass. Radiative decay is then modeled as a vertical transition within the Franck–Condon approximation, such that only the electronic state changes and the ground-state nuclear wavepacket immediately after decay is just a copy of the vibrational ground state in the excited-state PES, $\chi_g(q, t = 0) \propto \chi_e^{(0)}(q)$. The time evolution of this wavepacket on the ground-state PES thus follows the Schrödinger equation. We then analyze the probability (corresponding to the quantum yield) of finding the wavepacket in each isomer after a time $t_f \approx 160$ fs in which it has completely moved away from its initial configuration due to coherent motion. The quantum yield is then given by $\phi_{iso} = \int_{iso} |\Psi(q, t_f)|^2 dq$, where iso $\in \{A, B\}$ labels the isomer regions to the left and right of the energy barrier in $V_g(q)$. This gives a roughly equal reaction quantum yield for reaching either the stable (44%) or the metastable configuration (56%). As expected in a conventional photochemical reaction, the quantum yields in the bare molecule add up to one (indeed, the Stark-Einstein law can be reformulated as "the sum of quantum yields must be unity").

We note that the condition of high fluorescence quantum yield is not strictly necessary for the many-molecule reaction effects discussed below to take place. The important requirement is that the excited-state lifetime is sufficiently long, which precludes conical intersections between the ground and excited PES along or close to the reaction path. However, many molecules possess sloped conical intersections located at a higher energy along a nuclear coordinate orthogonal to q [45], where it is not easily reachable by the excited-state nuclear wavepacket. The associated bare-molecule nonradiative decay could be faster than radiative decay (which has typical timescales of nanoseconds), but would still be slow enough to allow the many-molecule reactions discussed here to take place. For simplicity, and to avoid having to introduce additional assumptions about the PES structure in directions orthogonal to the reaction coordinate q, we assume purely radiative decay in our calculations.

5.3.1 Single Molecule Quantum Yield Increase

We now consider a collection of these molecules placed inside a photonic structure supporting a single confined light mode. For simplicity, we assume perfect alignment between the molecular dipoles and the electric field direction. We again rely on the theoretical framework presented previously, through the use of PoPES resulting from diagonalization of the Hamiltonian in Eq. (4.4). We first study the scenario of $N = 5$ molecules, for which we show the coupled PES in Fig. 5.9a. Particularly, we present a

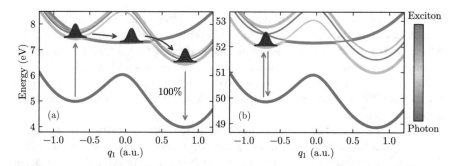

Fig. 5.9 Potential energy surfaces of a system with one light mode and **a** $N = 5$ molecules in the strong coupling regime with Rabi frequency $\Omega_R = 0.3$ eV and photon energy $\omega_c = 2.55$ eV, and **b** $N = 50$ molecules with Rabi frequency $\Omega_R = 0.75$ eV and photon energy $\omega_c = 2.4$ eV. All molecules but one are fixed at the initial position ($q_1 = q_{ms} \approx -0.7$ a.u.). The color scale represents the cavity mode fraction of the excited states, going from pure photon (purple) to pure exciton (orange). Partially adapted with permission from [44]. Copyright 2017 American Physical Society

cut of the five-dimensional PES where only the first molecule (q_1) is allowed to move, while all others are fixed to the equilibrium position of the metastable ground-state configuration ($q_i = q_{ms}$ for $i = 2, \ldots, 5$).

Already for the motion of just a single molecule, our results show that the quantum yield for the energy-releasing back-reaction can be significantly enhanced under strong coupling. The lowest-energy excited PoPES (see Fig. 5.9a) is formed by hybridization of the uncoupled excited-state surfaces of the molecules with the surface representing a photon in the cavity and the molecule in the ground state (a copy of the ground-state surface shifted upwards by the photon energy of the confined light mode). The photon energy ($\omega_c = 2.55$ eV) is close to resonant with the electronic excitation energy at the metastable configuration ($q = q_{ms}$), while most other molecular configurations (and specifically, the stable configuration $q = q_s$) are out of resonance with the cavity. This implies that the nature of the lowest excited-state PES changes depending on the molecular position q, corresponding to a polariton in some cases, and corresponding to a bare molecular state in others (as indicated by the usual color scale in Fig. 5.9). In the polaritonic states, each molecule is in its electronic ground state most of the time (since the excitation is distributed over all the molecules and the photonic mode), such that the hybrid parts of the lowest excited-state PoPES inherit their shape mostly from the ground-state PES, as discussed in Chap. 4. This leads to the formation of a new minimum in the lowest excited PoPES at the same position as the fully relaxed ground-state minimum q_s. The surface consists of two polaritonic regions (close to $q_1 = q_{ms}$ and $q_1 = q_s$) connected by an almost purely excitonic "bridge" (around $q_1 = 0$, where the cavity-exciton detuning is large), with smooth transitions between these parts. In the absence of barriers, a molecular system will quickly relax to the lowest-energy vibrational state on the lowest excited-state PoPES according to Kasha's rule [46]. Large-scale molecular dynamics calculations have recently shown that this rule also applies in polaritonic chemistry [47, 48].

Vibrational relaxation in the lowest excited hybrid light-matter PES will thus lead to localization of the nuclear wave packet close to the ground-state minimum q_s. As mentioned above, we assume that nonadiabatic couplings in the bare molecule are negligible along the reaction path, such that the dominant relaxation pathway is radiative decay. For the vibrationally relaxed wavepacket at $q_1 \approx q_s$, this would give a quantum yield of essentially unity for the back-reaction from the metastable to the stable configuration.

We note that this effect can be achieved because no energy barriers appear in the reaction path. Nevertheless, these can still emerge and thus the suppression effect discussed in Sect. 5.2 is possible with the adequate set of parameters. We show this in Fig. 5.9b, where we increased the number of molecules to $N = 50$, as well as increase the Rabi frequency and detune the photon energy. This introduces an energy barrier of $\approx$230 eV, which would lead to efficient trapping of the wavepacket, effectively suppressing the reaction. We thus find that precise tuning of the systems parameters can lead to two complete different outcomes.

5.3.2   Triggering of Many Reactions in Collective Strong Coupling

While the previous section already presented a large cavity-induced change of the photochemical properties of such molecules, we next show that the collective nature of the polaritons can result in even more dramatic qualitative changes in the system, allowing it to keep releasing energy during sequential relaxation of the molecules from the metastable to the stable configuration.

To understand this, we have to take into account that the PoPES formed under strong coupling encompass the nuclear degrees of freedom of *all* involved molecules. This collective nature can in particular also allow nuclear motion on *different* molecules to become coupled, and in the current case creates a reaction path along which the system can release the energy stored in all molecules, while staying on a single adiabatic PES reached by single-photon absorption in the initial state. This is demonstrated for motion of two of the involved molecules in Fig. 5.10a, which shows a two-dimensional cut of the PoPES of the lowest-energy excited state (with all other molecules again frozen in the metastable position $q = q_{ms}$). We calculate the minimum energy path (MEP) connecting the initial configuration $q_1 = q_2 = q_{ms}$ to the location where the first two molecules have released their stored energy ($q_1 = q_2 = q_s$) using the nudged elastic band method [49]. This approximate classical trajectory defines the reaction coordinate of the full "supermolecule" system. The initial position, for which we again assume that all molecules are at $q = q_{ms}$, corresponds to short-pulse excitation from the ground state in the metastable configuration, according to the Franck–Condon principle. The final position is $q_i = q_s$ for all i, i.e. the position where all the molecules are in the stable configuration (corresponding to the global minimum of the PES). It is worth noticing that due to the

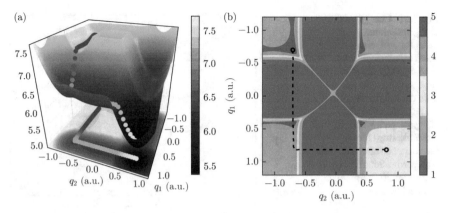

Fig. 5.10 **a** Lowest-energy excited state PES for 2 moving molecules in a 5-molecule ensemble. The minimum energy path (blue to white dots) connects the initial excited region with the final configuration of the two molecules. **b** Participation ratio map of the lowest-energy excited state, indicating over how many molecules the state is delocalized. The MEP is indicated by a dashed black line. Adapted with permission from [44]. Copyright 2017 American Physical Society

indistinguishability of our molecules, any of the available molecules can undergo the reaction in each step, and there are $N!$ equivalent paths from the initial to the final position. Due to rapid decoherence through interaction with the vibrational bath, we assume that quantum interference between these equivalent paths can be neglected, and we show only one of them in the following: the one in which the order of reactions corresponds to the numbering of the molecules. Along this path, indicated as a series of points in Fig. 5.10a, there are no significant reaction barriers, such that vibrational relaxation after absorption of a single photon indeed can lead to deactivation of both molecules. While we already proved in the previous section that simultaneous motion of several molecules is strongly suppressed in the low-excitation regime, we see that this does not prevent several reactions from occurring. The calculated MEP demonstrates that, to a good approximation, the reaction proceeds in steps, with the molecules moving one after the other (i.e., in the first leg, only q_1 changes, while in the second leg, only q_2 changes).

In order to gain additional insight into the properties of the polariton states that enable this step-wise many-molecule reaction triggered by a single photon, we further analyze the lowest excited PES by showing its molecular participation ratio in Fig. 5.10b. Here, the molecular participation ratio is defined as [50]

$$P_\alpha(\mathbf{q}) = \frac{\left(\sum_i |\langle e_i | \Psi_\alpha(\mathbf{q}) \rangle|^2\right)^2}{\sum_i |\langle e_i | \Psi_\alpha(\mathbf{q}) \rangle|^4}, \tag{5.6}$$

where $|e_i\rangle$ denotes the excited state of molecule i, and the sums are over all molecules. The participation ratio gives an estimate of the number of molecular states that possess a significant weight in a given state $|\Psi_\alpha\rangle$, with possible values ranging from $P_\alpha = 1$

Fig. 5.11 Energy profile
along the minimum energy
paths for collections of 2, 3,
4 and 50 molecules. For
$N = 50$, only the first five
steps are shown explicitly.
Thin dashed lines indicate
the approximate location
along the path where one
molecule stops moving and
the next one starts. Adapted
with permission from [44].
Copyright 2017 American
Physical Society

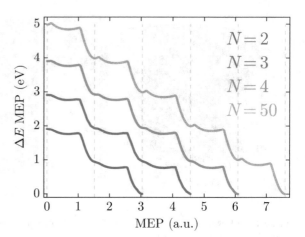

to $P_\alpha = N$ (for N molecules). Analyzing it for the lowest-energy excited state PoPES (see Fig. 5.10b) demonstrates that the surface at the starting point corresponds to a collective polariton, with the excitation equally distributed over all molecules. Along the MEP, the excitation collapses onto a single molecule (the one that is moving), demonstrated by the participation ratio decreasing to 1 for -0.5 a.u. $\lesssim q_1 \lesssim 0.4$ a.u.. As the molecule moves, it again enters into resonance with the cavity (and the other molecules) and the state changes character to a fully delocalized polariton with $P_{LP} = N$ (at $q_1 \approx 0.45$ a.u.). However, as the first molecule keeps moving, it falls out of resonance again and effectively "drops out" of the polaritonic state, leaving the excitation in a polaritonic state distributed over the photonic mode and the remaining $N - 1$ molecules ($P_{LP} = 4$), which then forms the starting point for the second molecule to undergo the reaction. Following the MEP along the second leg (where $q_1 \approx q_s$ and q_2 moves from q_{ms} to q_s), the same process repeats, but now involving one less molecule.

We now demonstrate that the same process can keep repeating for many molecules. To this end, we calculate the MEP for varying numbers of molecules from $N = 2$ to $N = 50$, with a collective Rabi splitting of $\Omega_R = 0.3$ eV in the initial molecular configuration ($q_i = q_{ms}$ for all i) for all cases. As shown in Fig. 5.11, the energy profile along the MEP is structurally similar for any number of molecules. The main change is that for larger values of N, the collective protection effect makes the PES resemble the shape of the uncoupled PES more strongly, leading to a less smooth MEP with slightly higher barriers, comparable to the average thermal kinetic energy at room temperature. In addition, the significant change of collective state when passing the barrier (with the excitation collapsing from all molecules onto a single one in the "bridge" region around $q = 0$) leads to narrow avoided crossings in the adiabatic picture. As in previous chapters, in a diabatic picture formed by the polaritonic PES of $N - 1$ coupled molecules and the remaining bare excited PES, their coupling can be shown to be proportional to the single-molecule coupling strength $\Omega_R/\sqrt{N}$, such that to lowest order in perturbation theory, the transition probability from the polaritonic to the pure-exciton surface scales as $1/N$. However, this effect is compensated by the fact that there are many possible equivalent paths, corresponding to motion of any of the remaining metastable molecules. We thus again assume that TST for a single barrier of the same height provides a reasonable estimate for the average time needed to overcome any one of the barriers. This gives $\tau \approx \frac{h}{k_B T} \exp\left(\frac{\Delta E}{k_B T}\right) \lesssim 1$ ps at room temperature [51]. It should be noted that several other mechanisms relevant in the current case imply that the TST predictions correspond to an upper limit, as TST is known to fail in multi-step reactions where the nuclear wavepacket approaches the barriers with some initial kinetic energy [52], and also neglects quantum tunneling effects that are important for small energy barriers as found here.

The estimated times for passing the barriers highlight the importance of the lifetime of the hybrid light–matter states to determine the feasibility of triggering multiple reactions with a single photon. In most current experiments, polariton lifetimes (which are an average of the lifetimes of their constituents) are very short, on the order of tens of femtoseconds, due to the use of short-lived photonic modes such as localized surface plasmons or low-Q modes in metallic and dielectric Fabry–Perot microcavities. In contrast, the lifetime of the molecular excitations can be limited by their spontaneous radiative decay, which is on the order of nanoseconds for typical organic molecules. Consequently, if long-lived photonic modes as available in low-loss dielectric structures such as photonic crystals or microtoroidal cavities are used instead, there is no fundamental reason preventing polariton lifetimes that approach nanoseconds. This would thus give enough time for thousands of molecules to undergo a reaction before the excitation is lost due to radiative decay.

5.4 Conclusions

To conclude this chapter, we have explored two different possibilities of manipulating photochemistry using light–matter strong coupling. We first have demonstrated the stabilization of excited-state molecular structure and accompanying strong suppression of photochemical reactions under strong coupling of molecules to confined light modes. While already effective in the case of a single coupled molecule, we find that collective coupling of a large number of molecules to a single light mode provides an even stronger stabilization due to the collective protection effect. We additionally find that this phenomenon does not vanish in higher excited subspaces. These results do not depend on the specifics of the molecular model, such that the observed stabilization is expected to occur for any kind of photochemical reaction that is induced by motion on the excited molecular PES.

The combination of the effects discussed in Sect. 5.2 leads to an almost complete suppression of photoisomerization, with a potentially much larger predicted effect than the change in rate observed in [1]. There are two main reasons for this difference. First, in the experiment the isomer representing the product was the one in strong coupling, unlike in our study, meaning that the molecule was much less affected by the coupling to the cavity. Second, in here we treat a single confined light mode while the experiment consist of a planar microcavity that hosts a continuum of light modes. However, a more similar setup was achieved in a 2016 study by the group of Timur Shegai, where they showed experimentally the possibility of a 100-fold reduction of the rate of photo-oxidation of organic dyes by strongly coupling them to plasmonic nanoantennas [2]. This has been further demonstrated in early 2019, where photodegradation of the semiconducting polymer P3HT has been reduced threefold [3]. This confirms that the energy landscape is sufficiently altered in the strong coupling regime to strongly influence the reaction kinetics.

Finally, in Sect. 5.3, we have demonstrated that under strong coupling, a single photon could be used to trigger a photochemical reaction in *many* molecules. This corresponds to an effective quantum yield (number of reactant molecules per absorbed photon) of the reaction that is significantly larger than one, and thus provides a possible pathway to break the second (or Stark–Einstein) law of photochemistry without relying on fine-tuned resonance conditions. The basic physical effect responsible for this surprising feature is the delocalized nature of the polaritonic states obtained under collective strong coupling, which require a treatment of the whole collection of molecules as a single polaritonic "supermolecule". For the specific model studied here, this strategy could resolve one of the main problems of solar energy storage with organic molecules: How to efficiently retrieve the stored energy from molecules that are designed for the opposite purpose, i.e., for storing energy very efficiently under normal conditions [42, 43]. By reversibly bringing the system into strong coupling (e.g., through a moving mirror that brings the cavity into and out of resonance), one could thus trigger the release of the stored energy through absorption of a single ambient photon.

References

1. Hutchison JA, Schwartz T, Genet C, Devaux E, Ebbesen TW (2012) Modifying chemical landscapes by coupling to vacuum fields. Angew Chem Int Ed 51:1592
2. Munkhbat B, Wersäll M, Baranov DG, Antosiewicz TJ, Shegai T. Suppression of photo-oxidation of organic chromophores by strong coupling to plasmonic nanoantennas
3. Peters VN, Faruk MO, Asane J, Alexander R, D'angelo AP, Prayakarao S, Rout S, Noginov M (2019) Effect of strong coupling on photodegradation of the semiconducting polymer P3HT. Optica 6:318
4. Polli D, Altoè P, Weingart O, Spillane KM, Manzoni C, Brida D, Tomasello G, Orlandi G, Kukura P, Mathies RA, Garavelli M, Cerullo G (2010) Conical intersection dynamics of the primary photoisomerization event in vision. Nature 467:440
5. Kucharski TJ, Tian Y, Akbulatov S, Boulatov R (2011) Chemical solutions for the closed-cycle storage of solar energy. Energy Environ Sci 4:4449
6. Irie M, Fukaminato T, Matsuda K, Kobatake S (2014) Photochromism of diarylethene molecules and crystals: memories, switches, and actuators. Chem Rev 114:12174
7. Guentner M, Schildhauer M, Thumser S, Mayer P, Stephenson D, Mayer PJ, Dube H (2015) Sunlight-powered kHz rotation of a hemithioindigo-based molecular motor. Nat Commun 6:8406

8. Zietz B, Gabrielsson E, Johansson V, El-Zohry AM, Sun L, Kloo L (2014) Photoisomerization of the cyanoacrylic acid acceptor group—a potential problem for organic dyes in solar cells. Phys Chem Chem Phys 16:2251
9. Sinha RP, Häder D-P (2002) UV-induced DNA damage and repair: a review. Photochem Photobiol Sci 1:225
10. Douki T, Reynaud-Angelin A, Cadet J, Sage E (2003) Bipyrimidine photoproducts rather than oxidative lesions are the main type of DNA damage involved in the genotoxic effect of solar UVA radiation. Biochemistry 42:9221
11. Waldeck DH (1991) Photoisomerization dynamics of stilbenes. Chem Rev 91:415
12. Quick M, Dobryakov AL, Gerecke M, Richter C, Berndt F, Ioffe IN, Granovsky AA, Mahrwald R, Ernsting NP, Kovalenko SA (2014) Photoisomerization dynamics and pathways of trans- and cis-azobenzene in solution from broadband femtosecond spectroscopies and calculations. J Phys Chem B 118:8756
13. Bonačić-Koutecký V, Bruckmann P, Hiberty P, Koutecký J, Leforestier C, Salem L (1975) Sudden polarization in the zwitterionic Z 1 excited states of organic intermediates. Photochemical implications. Angew Chem Int Ed English 14:575
14. Schneider BI, Feist J, Nagele S, Pazourek R, Hu S, Collins LA, Burgdörfer J (2011) Recent advances in computational methods for the solution of the time-dependent Schrödinger equation for the interaction of short, intense radiation with one and two-electron systems: application to He and H_2^+. In: Bandrauk AD, Ivanov M (eds) Quantum dynamic imaging. CRM series in mathematical physics, vol 149. Springer, New York, NY
15. Feist J (2019) Collection of small useful helper tools for Python. https://github.com/jfeist/jftools
16. Litinskaya M, Reineker P, Agranovich VM (2004) Fast polariton relaxation in strongly coupled organic microcavities. J Lumin 110:364
17. Coles DM, Michetti P, Clark C, Adawi AM, Lidzey DG (2011) Temperature dependence of the upper-branch polariton population in an organic semiconductor microcavity. Phys Rev B 84:205214
18. Chikkaraddy R, de Nijs B, Benz F, Barrow SJ, Scherman OA, Rosta E, Demetriadou A, Fox P, Hess O, Baumberg JJ (2016) Single-molecule strong coupling at room temperature in plasmonic nanocavities. Nature 535:127
19. Wiederrecht GP, Wurtz GA, Hranisavljevic J (2004) Coherent coupling of molecular excitons to electronic polarizations of noble metal nanoparticles. Nano Lett 4:2121
20. Zengin G, Wersäll M, Nilsson S, Antosiewicz TJ, Käll M, Shegai T (2015) Realizing strong light-matter interactions between single-nanoparticle plasmons and molecular excitons at ambient conditions. Phys Rev Lett 114:157401
21. Galego J, Garcia-Vidal FJ, Feist J (2015) Cavity-induced modifications of molecular structure in the strong-coupling regime. Physical Review X 5:41022
22. Ćwik JA, Kirton P, De Liberato S, Keeling J (2016) Excitonic spectral features in strongly coupled organic polaritons. Phys Rev A 93:033840
23. Feist J, Galego J, Garcia-Vidal FJ (2018) Polaritonic chemistry with organic molecules. ACS Photonics 5:205
24. Virgili T, Coles D, Adawi A, Clark C, Michetti P, Rajendran S, Brida D, Polli D, Cerullo G, Lidzey D (2011) Ultrafast polariton relaxation dynamics in an organic semiconductor microcavity. Phys Rev B 83:2
25. Barachati F, De Liberato S, Kéna-Cohen S (2015) Generation of Rabi-frequency radiation using exciton-polaritons. Phys Rev A 92:033828
26. Gubbin CR, Maier SA, De Liberato S (2017) Theoretical investigation of phonon polaritons in SiC micropillar resonators. Phys Rev B 95:035313
27. Barachati F, Simon J, Getmanenko YA, Barlow S, Marder SR, Kéna-Cohen S (2017) Tunable third-harmonic generation from polaritons in the ultrastrong coupling regime. ACS Photonics 5:119
28. Kéna-Cohen S, Forrest SR (2010) Room-temperature polariton lasing in an organic single-crystal microcavity. Nat Photonics 4:371

29. Daskalakis KS, Maier SA, Murray R, Kéna-Cohen S (2014) Nonlinear interactions in an organic polariton condensate. Nat Mater 13:271
30. Plumhof JD, Stöferle T, Mai L, Scherf U, Mahrt RF (2014) Room-temperature Bose-Einstein condensation of cavity exciton-polaritons in a polymer. Nat Mater 13:247
31. Rodriguez SRK, Feist J, Verschuuren MA, Garcia Vidal FJ, Gómez Rivas J (2013) Thermal- ization and cooling of plasmon-exciton polaritons: towards quantum condensation. Phys Rev Lett 111
32. Ramezani M, Halpin A, Fernández-Domínguez AI, Feist J, Rodriguez SR-K, Garcia-Vidal FJ, Gómez Rivas J (2017) Plasmon-exciton-polariton lasing. Optica 4:31
33. Holstein T, Primakoff H (1940) Field dependence of the intrinsic domain magnetization of a ferromagnet. Phys Rev 58:1098
34. Sáez-Blázquez R, Feist J, Fernández-Domínguez A, García-Vidal F (2017) Enhancing photon correlations through plasmonic strong coupling. Optica 4:1363
35. Sáez-Blázquez R, Feist J, García-Vidal F, Fernández-Domínguez A (2018) Photon statistics in collective strong coupling: nanocavities and microcavities. Phys Rev A 98:013839
36. Rohatgi-Mukherjee KK (2013) Fundamentals of photochemistry. New Age International
37. Summers DP, Luong JC, Wrighton MS (1981) A new mechanism for photosubstitution of organometallic complexes. Generation of substitutionally labile oxidation states by excited- state electron transfer in the presence of ligands. J Am Chem Soc 103:5238
38. Arai T, Karatsu T, Sakuragi H, Tokumari K (1983) One-way photoisomerization between cis- and trans-olefin. A novel adiabatic process in the excited state. Tetrahedron Lett 24:2873
39. Eves BJ, Sun Q-Y, Lopinski GP, Zuilhof H (2004) Photochemical attachment of organic mono- layers onto H-terminated Si(111): radical chain propagation observed via STM studies. J Am Chem Soc 126:14318
40. Walker BJ, Musser AJ, Beljonne D, Friend RH (2013) Singlet exciton fission in solution. Nat Chem 5:1019
41. Zirzlmeier J, Lehnherr D, Coto PB, Chernick ET, Casillas R, Basel BS, Thoss M, Tykwinski RR, Guldi DM (2015) Singlet fission in pentacene dimers. Proc Natl Acad Sci 112:5325
42. Cacciarini M, Skov AB, Jevric M, Hansen AS, Elm J, Kjaergaard HG, Mikkelsen KV, Brønd- sted Nielsen M (2015) Towards solar energy storage in the photochromic dihydroazulene- vinylheptafulvene system. Chem Eur J 21:7454
43. Gurke J, Quick M, Ernsting NP, Hecht S (2017) Acid-catalysed thermal cycloreversion of a diarylethene: a potential way for triggered release of stored light energy? Chem Commun 53:2150
44. Galego J, Garcia-Vidal FJ, Feist J (2017) Many-molecule reaction triggered by a single photon in polaritonic chemistry. Phys Rev Lett 119:136001
45. Levine BG, Martínez TJ (2007) Isomerization through conical intersections. Annu Rev Phys Chem 58:613
46. Kasha M (1950) Characterization of electronic transitions in complex molecules. Discuss Faraday Soc 9:14
47. Baieva S, Hakamaa O, Groenhof G, Heikkila TT, Toppari JJ (2017) Dynamics of strongly coupled modes between surface plasmon polaritons and photoactive molecules: the effect of the stokes shift. ACS Photonics 4:28
48. Luk HL, Feist J, Toppari JJ, Groenhof G (2017) Multiscale molecular dynamics simulations of polaritonic chemistry. J Chem Theory Comput 13:4324
49. Henkelman G, Uberuaga BP, Jónsson H (2000) A climbing image nudged elastic band method for finding saddle points and minimum energy paths. J Chem Phys 113:9901
50. Kramer B, MacKinnon A (1993) Localization: theory and experiment. Rep Prog Phys 56:1469
51. Eyring H (1935) The activated complex in chemical reactions. J Chem Phys 3:107
52. Anslyn EV, Dougherty DA (2006) Modern physical organic chemistry. University Science Books

Chapter 6
Cavity Ground-State Chemistry

6.1 Introduction

In recent years, the possibility of influencing the thermally driven reactivity of organic molecules in the electronic *ground* state has been demonstrated by coupling the cavity to vibrational transitions of the molecules [1–4]. This opens a wide range of possibilities, such as cavity-enabled catalysis and manipulation of ground-state chemical processes. In this chapter we theoretically investigate the possibility of modifying ground-state chemical properties of organic molecules. Other attempts to understanding these experimental observations have been done. More specifically, it has been shown that chemical reactions are not strongly modified even under electronic ultrastrong collective coupling [5, 6]. Additionally, in a series of papers based on more microscopic models, Flick and co-workers have shown that ground state properties can be significantly modified under single-molecule (ultra-)strong coupling [7–9], but have not explicitly treated chemical reactivity. It has also been reported resonant enhancement of ground-state electron transfer reactions in more specific theoretical descriptions [10].

In the present chapter, we aim to understand cavity-induced modifications of ground-state chemistry in coupled molecule-cavity systems using a general theoretical model. In Sect. 6.2 we present the light–matter interaction Hamiltonian for a single molecule coupled to a nanoscale cavity. Then, we review the cavity Born–Oppenheimer approximation [8, 9], which allows to approach polaritonic chemistry by treating the photonic DoF as a continuous parameter and on equal footing as the nuclear DoF. We then present the Shin–Metiu model [11], a simple molecular model that displays a possible chemical reaction and that allows us to perform calculations with the full Hamiltonian without invoking any approximation. Then in Sect. 6.3 we start by obtaining the formally exact quantum reaction rates for the system [12–14]. In order to understand these results we develop a simplified theory based on the cavity Born–Oppenheimer approximation [8] and on perturbation theory, where we

© The Editor(s) (if applicable) and The Author(s), under exclusive license
to Springer Nature Switzerland AG 2020
J. Galego Pascual, *Polaritonic Chemistry*, Springer Theses,
https://doi.org/10.1007/978-3-030-48698-3_6

find that we can predict the reaction rate changes based on transition state theory [15, 16]. Furthermore, this theory allows us to make explicit connections to electrostatic, van der Waals, and Casimir–Polder interactions. In Sect. 6.4 we present two different calculations in realistic systems such as a nanoparticle-on-mirror cavity [17–19], where the single-molecule coupling can be significant, and on the change in rate in the internal rotation of the 1,2-dichloroethane molecule, where we demonstrate the full potential of the cavity to inhibit or catalyze reactions, or even to modify the equilibrium configuration of the molecule. Then in Sect. 6.5 we extend our model to an ensemble of molecules and find collective enhancement of the effect under orientational alignment of the molecular dipoles. We additionally discuss collective phenomena in Sect. 6.6, where the change of the ground-state equilibrium structure of the molecule is investigated, also using an approach of polaritonic potential energy surfaces.

We mention here that we do not explicitly treat the case of many molecules coupled to a cavity with a continuum of modes, i.e., the case which corresponds to the experimentally used Fabry–Perot cavities with in-plane dispersion [1, 4]. For the sake of simplicity, we also neglect solvent effects. While these are well-known to be important in chemical reactions, their effect depends strongly on the chosen solvent and experimental setup (particularly in nanocavities). However, we mention that the latest experimental studies indicate that solvent effects might be responsible and/or relevant for the experimentally observed resonance-dependent ground-state chemical reactivity [2, 3].

6.2 Theoretical Model

We restrict the following discussion to organic molecules coupled to a nanocavity, based on the Hamiltonian within the quasistatic approximation presented in Sect. 2.3. For simplicity, we first consider a single molecule including n_e electrons and n_n nuclei. The Hamiltonian is thus

$$\hat{H} = \sum_{i=1}^{n_n} \frac{\hat{\mathbf{P}}_i^2}{2M_i} + \hat{H}_e(\hat{\mathbf{x}}, \hat{\mathbf{R}}) + \sum_k \omega_k \left(\hat{a}_k^\dagger \hat{a}_k + \frac{1}{2} \right) + \sum_k \omega_k \hat{q}_k \boldsymbol{\lambda}_k \cdot \hat{\boldsymbol{\mu}}(\hat{\mathbf{x}}, \hat{\mathbf{R}}). \quad (6.1)$$

The bare molecular Hamiltonian corresponds to the first two terms: the kinetic energy of n_n nuclei and the electronic Hamiltonian. The latter includes the kinetic energy of the n_e electrons and the nucleus–nucleus, electron–electron, and nucleus–electron interaction potentials. This operator depends on all the electronic and nuclear positions, $\hat{\mathbf{x}} = (\hat{\mathbf{x}}_1, \hat{\mathbf{x}}_2, \ldots, \hat{\mathbf{x}}_{n_e})$ and $\hat{\mathbf{R}} = (\hat{\mathbf{R}}_1, \hat{\mathbf{R}}_2, \ldots, \hat{\mathbf{R}}_{n_n})$, respectively. In Eq. (6.1) we now use the photonic displacement[1] $\hat{q}_k = \frac{1}{\sqrt{2\omega_k}}(\hat{a}_k^\dagger + \hat{a}_k)$ and the electric field strength is determined by the coupling strength $\boldsymbol{\lambda}_k = \lambda_k \boldsymbol{\epsilon}_k$. This

---

[1]Do not confuse the photonic displacement q_k with the generalized nuclear coordinate $\mathbf{q}$ used in previous chapters. To avoid confusion, in this chapter nuclear coordinates are explicitly denoted $\mathbf{R}$.

coupling constant can be related to both the single-mode electric field strength and the (position-dependent) effective mode volume of the quantized mode, with $\lambda_k = \sqrt{\frac{2}{\omega_k}} E_{1ph,k}(\mathbf{r}_m) = \sqrt{4\pi/V_{eff,k}}$. Here we use the general definition of the effective EM mode volume of Eq. (2.71).

In the following, we will first treat a cavity in which only a single mode has significant coupling to the molecule. Since the interaction depends on the inner product between the electric field and the total dipole moment $\hat{\boldsymbol{\mu}} = \sum_i^{n_n} Z_i \hat{\mathbf{R}}_i - \sum_i^{n_e} \hat{\mathbf{x}}_i$, only the projection $\hat{\mu}_\epsilon = \hat{\boldsymbol{\epsilon}} \cdot \hat{\boldsymbol{\mu}}$ is relevant, and we only have to deal with scalar quantities. For the sake of simplicity, we rewrite $\hat{\mu}_\epsilon \to \hat{\mu}$. We also assume perfect alignment between the molecule and the field unless indicated otherwise.

6.2.1 Cavity Born–Oppenheimer Approximation

In order to treat molecules coupled to low-energy photons (such as in vibrational strong coupling) we make use of the cavity Born–Oppenheimer approximation [9]. We now review in detail this description, which starts by expressing the photonic DoF as an explicit harmonic oscillator, where the electromagnetic energy of the k-th mode reads

$$\hat{H}_{EM}^{(k)} = \omega_k \left(\hat{a}_k^\dagger \hat{a}_k + \frac{1}{2} \right) = \frac{\hat{p}_k^2}{2} + \omega_k^2 \frac{\hat{q}_k^2}{2}, \tag{6.2}$$

with $\hat{p}_k = i\sqrt{\omega_k/2} \left(\hat{a}_k^\dagger + \hat{a}_k \right)$ and $\hat{q}_k = 1/\sqrt{2\omega_k} \left(\hat{a}_k^\dagger + \hat{a}_k \right)$ as the photon canonical momentum and displacement respectively. By comparing this to the electromagnetic energy presented in Sect. 2.1, we see that the photon displacement is directly related to the electric displacement field associated to that mode through $\hat{\mathbf{D}}_k = \omega_k \lambda_k \hat{q}_k^2$ and the photon momentum is related to the magnetic field [20, 21].

Within the explicit harmonic oscillator description for the electromagnetic Hamiltonian, it is possible now to perform an adiabatic separation similar to the standard BOA. This way of writing the photonic Hamiltonian using a continuous photonic displacement operator suggests a different approach for treating the photon modes than employed in the previous chapters: by treating the photonic DoF on equal footing to a nuclear coordinate within the BOA. Therefore, we include the photonic DoF in the Born–Huang expansion (see Sect. 2.2):

$$\Psi(\mathbf{r}_i, \mathbf{R}_j, q_k) = \sum_l \chi_l(\mathbf{R}_j, q_k) \Phi_l(\mathbf{r}_i; \mathbf{R}_j, q_k), \tag{6.3}$$

where for simplicity we treat the single-molecule case, and thus $\mathbf{r}_i$, $\mathbf{R}_j$, and q_k represent all the electronic and nuclear DoF of the molecule, and photonic coordinates

---

[2]Note that the vector dependence is in the coupling constant $\lambda_k = \lambda_k \mathbf{e}_k$ so that the photonic displacement $\hat{q}_k$ is a scalar operator.

of all the modes, respectively. The electronic eigenstates satisfy

$$\hat{H}_e(\hat{\mathbf{r}}_i; \mathbf{R}_j, q_k)\Phi_l(\mathbf{r}_i; \mathbf{R}_j, q_k) = V_l(\mathbf{R}_j, q_k)\Phi_l(\mathbf{r}_i; \mathbf{R}_j, q_k), \qquad (6.4)$$

where the electronic Hamiltonian is now

$$\hat{H}_e(\hat{\mathbf{r}}_i; \mathbf{R}_j, q_k) = \hat{H}_{tot} - \hat{T}_n - \sum_k \frac{\hat{p}_k^2}{2} = \hat{H}_{mol}(\hat{\mathbf{r}}_i; \mathbf{R}_j) + \sum_k \left(\frac{\omega_k^2}{2} q_k^2 + \omega_k \hat{q}_k \boldsymbol{\lambda}_k \cdot \hat{\boldsymbol{\mu}}(\hat{\mathbf{r}}_i; \mathbf{R}_j) \right).$$
$$(6.5)$$

In this case the electronic wavefunctions depend parametrically on both the nuclear configuration and the photonic displacements. By replacing this in the Schrödinger equation we again find a set of differential equations similar to the ones described in Eq. (2.29) of Sect. 2.1

$$\left[\hat{T}_n + \frac{\hat{p}_k^2}{2} + V_l(\mathbf{R}_j, q_k) \right] \chi_l(\mathbf{R}_j, q_k) + \sum_{l'} \hat{\Lambda}_{ll'}(\mathbf{R}_j, q_k)\chi_{l'}(\mathbf{R}_j, q_k) = E\chi_l(\mathbf{R}_j, q_k),$$
$$(6.6)$$

which are coupled through the new nonadiabatic term

$$\hat{\Lambda}_{ll'}^{(cav)}(\mathbf{R}_j, q_k) = \langle \Phi_l(\mathbf{r}_i; \mathbf{R}_j, q_k)| \left(\hat{T}_n + \frac{\hat{p}_k^2}{2} \right) |\Phi_{l'}(\mathbf{r}_i; \mathbf{R}_j, q_k)\rangle_{\mathbf{r}_i} - \left(\hat{T}_n + \frac{\hat{p}_k^2}{2} \right) \delta_{ll'}.$$
$$(6.7)$$

The *cavity Born–Oppenheimer approximation* (CBOA) [8, 9] consists on neglecting these nonadiabatic terms, and thus considering the electronic PES (now dependent on nuclear and photonic degrees of freedom) completely independent. Due to the formal equivalence between nuclear and photonic degrees of freedom within this picture,[3] all the results and standard procedure discussed in Sect. 2.2 can be extended to the CBOA. Indeed, the emergent nonadiabatic terms will become relevant when the electronic PES are close in energy. As we later discuss, the separability condition is precisely fulfilled in vibrational strong coupling, where photonic and nuclear excitation energies are comparable [9].

Note that this picture does not give any precise insight of the level of hybridization of light and matter. In conventional strong coupling this is achieved through the Rabi splitting relative to the resonance frequency Ω_R/ω_0, where Ω_R is the energy separation between polaritons. In this picture, in order to obtain Ω_R, it is required to explicitly calculate the coupled nuclear-photonic eigenstates determined by the new cavity-PESs. This is most easily achieved close to local minima, where the surface can be approximated through coupled harmonic oscillator potentials. A standard procedure in chemistry is to diagonalize the corresponding Hessian of the surface to obtain the new normal modes. Let us thus analyze the case of vibrational SC in the ground-state. Consider that once the CBOA has been performed for a system with

---

[3]Note that both Eqs. 2.29 and 6.6 represent the same equation if we considered the photonic term $\frac{\hat{p}_k^2}{2}$ as the kinetic energy of another nuclear DoF.

one nuclear degree of freedom and one photonic mode, the ground-state PES close
to the minimum is given by

$$V_g(R, q) = \frac{\omega_\nu^2}{2} R^2 + \frac{\omega_c^2}{2} q^2 + \lambda \omega_c q \mu_g(R),$$ (6.8)

where we are here using mass-weighted coordinates ($R \to R/\sqrt{M}$) for the nuclear
coordinate of vibrational frequency ω_ν and $\mu_g(R) = \langle g(R)|\hat{\mu}|g(R)\rangle$ is the ground-
state permanent dipole moment. The Hessian of the surface is

$$\mathcal{H} = \begin{pmatrix} \omega_\nu^2 & \lambda \omega_c \mu_g'(R_0) \\ \lambda \omega_c \mu_g'(R_0) & \omega_c^2 \end{pmatrix},$$ (6.9)

where $\mu_g'(R_0)$ is the derivative of the ground-state dipole moment evaluated at
the minimum $R_0 = 0$. The eigenvalues of the Hessian correspond to the squares
of the normal modes frequencies. In the resonant case with the first vibrational
frequency ($\omega_c = \omega_\nu$) it is straightforward to show that the new frequencies are
$\omega_\pm = \omega_c \sqrt{1 \pm \frac{\lambda}{\omega_c} \mu_g'(R_0)}$. This is the standard result for the modes of two coupled
harmonic oscillators beyond the RWA [22]. The connection between the coupling
strength and the Rabi splitting is clearer in the low-coupling limit:

$$\omega_\pm \approx \omega_c \pm \frac{1}{2} \lambda \mu_g'(R_0).$$ (6.10)

The Rabi splitting to lowest order is then $\Omega_R = \lambda \mu_g'(R_0)$, i.e., proportional to λ.
The first derivative of the momentum corresponds to the transition dipole moment
between nuclear eigenstates, making thus the connection to the Rabi splitting presented in previous sections. This derivation is equivalent to the one performed in
[23], where they demonstrated vibrational strong coupling in organic molecules for
the first time.

6.2.2 Shin–Metiu Model

In order to study changes in ground-state chemical reactivity induced by strong
coupling to a nanocavity, we first treat a simple molecular model system that is
numerically fully solvable and has been extensively studied in model calculations
of chemical reaction rates, the Shin–Metiu model [11]. It treats three nuclei and
one electron moving in one dimension, as presented in Fig. 6.1a. Two of the nuclei
are separated by a distance L and fixed in place, while the remaining nucleus and
the electron are free to move. The repulsive interaction of the mobile nucleus with
the fixed ones is given by a normal Coulomb potential, while the attractive electron–
nuclei interaction is given by softened Coulomb potentials $V_{en}(r_i) = Z \mathrm{erf}(r_i/R_c)/r_i$,

where r_i is the distance between the electron and nucleus i and R_c is the soften-ing parameter. The system has two stable nuclear configurations (minima of the ground-state Born–Oppenheimer surface) that represent two different isomers of a charge or proton transfer reaction. Given that the electronic excitations energies and thus the nonadiabatic couplings between different potential energy surfaces can be varied easily by changing the parameters of the Shin–Metiu model, it has been extensively studied in the context of correlated electron–nuclear dynamics [24, 25], as well as in the context of polariton formation under strong coupling [8, 9]. The parameters chosen throughout the present work are $Z = 1$, $L = 10$ Å ≈ 18.9 a.u., $M = 1836$ a.u., and $R_c = 1.5$ Å ≈ 2.83 a.u. (for all three nuclei), resulting in the Born–Oppenheimer potential energy surfaces shown in Fig. 6.1b, with negligible nonadiabatic coupling between electronic surfaces. The figure also shows the first few vibrational eigenstates close to each minimum (tunneling through the central energy barrier is negligible for these states, so that they can be chosen to be localized on the left or right, respectively). In Fig. 6.1c we show the ground-state permanent dipole moment $\mu_g(R) = \langle g | \mu(R) | g \rangle$. Below we demonstrate that, to a good approxi-

Fig. 6.1 **a** Schematic representation of the Shin–Metiu model close to one of the equilibrium configurations. The two ions on both sides are fixed at a distance L, while the electron and the remaining ion can move freely in between. **b** Potential energy surfaces of the model with the vibrational levels and associated probability densities of the ground state (blue) represented. **c** Ground state dipole moment

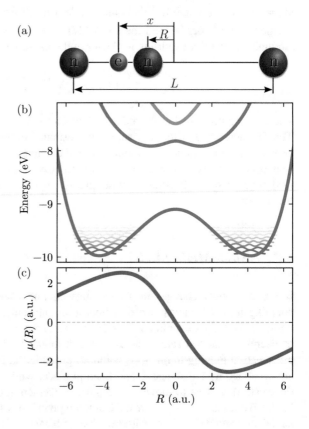

mation, the ground-state potential energy surface and dipole moment are sufficient to describe the change in the molecular ground-state structure and chemical reactivity due to the cavity.

6.3 Effects of the Cavity on Ground-State Reactivity

In this section we study the changes on the ground-state reactivity of the Shin–Metiu model induced by a single cavity mode, using the Hamiltonian Eq. (6.1). We first analyze how the reaction rates are modified when increasing the light–matter coupling λ. We then develop a theory in which we turn to the cavity Born–Oppenheimer approximation and perturbation theory to explain and predict possible changes.

6.3.1 Reaction Rates

The Shin–Metiu model presents a possible ground-state proton-transfer reaction from the left minimum at $R \approx 4$ a.u. to the right one (or vice versa). In the following, we take advantage of the simplicity of the Shin–Metiu model to exactly compute the quantum reaction rates without any approximation, as reviewed in Sect. 2.2. This automatically takes into account all quantum effects, including tunneling and zero-point energy. In particular, we will find the reaction rates by using Eq. (2.42), which for convenience we rewrite here:

$$k(T) = \frac{1}{Q_r(T)} \int_0^{t_f \to \infty} C_{ff}(t) dt. \tag{6.11}$$

In order to obtain the flux operator $\bar{F} = \frac{1}{2M}(\hat{P}s'(R)\delta(s) + \delta(s)\hat{P}s'(R))$, required to calculate $C_{ff}(t)$, we define the dividing surface as $s(R) = R$, such that it divides reactants and products at $R = 0$.

In order to obtain the rates of the coupled electronic-nuclear-photonic system, we discretize all three degrees of freedom, using a finite-element discrete variable representation [26, 27] for x and R, as well as the Fock basis for the cavity photon mode. This allows to diagonalize the full Hamiltonian and thus to straightforwardly calculate the flux-flux autocorrelation function Eq. (2.43) for arbitrary time t. For numerical efficiency, we perform the diagonalization in steps, first diagonalizing the bare molecular Hamiltonian, performing a cut-off in energy, and then diagonalizing the coupled system in this basis. We have carefully checked convergence with respect to all involved grid and basis set parameters and cutoffs. As is well known [11, 24], due to the absence of dissipation in the model, for large times the correlation function becomes negative and oscillates around zero, corresponding to the wave packet that has crossed the barrier returning back through the dividing surface after reflection at the other side of the potential (at $R \approx 6$ a.u.). However, in a real system the reaction coordinate is coupled to other vibrational and solvent degrees of freedom that will dissipate the energy and prevent recrossing. To represent this, we choose a final time t_f around which the correlation function stays equal to zero for a while and only integrate up that time in Eq. (6.11). The time chosen, $t_f = 35$ fs, corresponds to typical dissipation times in condensed phase reactions, and is similar to values chosen in the cavity-free case [11].

We now study the cavity-modified chemical reaction rates of the hybrid system for different coupling strengths λ. In order to evaluate the strength of the coupling, we note here that at the resonance condition with vibrational transitions we can link the coupling constant λ with the Rabi splitting Ω_R. Note that in the formation of vibro-polaritons, i.e., hybridization of the photon mode with the vibrational transitions of the molecule, the Rabi splitting is determined by the transition dipole moment and frequency of the quantized vibrational levels of the molecule. Within a lowest-order expansion around the equilibrium position (see Eq. (6.10) and related discussion), $V_g(R) \approx V_g(R_0) + \frac{1}{2} M \omega_\nu^2 (R - R_0)^2$, $\mu_g(R) \approx \mu_g(R_0) + \mu_g'(R_0)(R - R_0)$, these are given by $\omega_\nu = 72.6$ meV, and $\mu_\nu \approx \frac{1}{\sqrt{2M\omega_\nu}} \mu_g'(R_0)$, giving a Rabi frequency $\Omega_R = \frac{\lambda}{\sqrt{M}} \mu_g'(R_0)$ on resonance ($\omega_c = \omega_\nu$) [23]. We note that a coupling strength of $\lambda = 0.035$ a.u. corresponds to a Rabi splitting of $\Omega_R \approx 0.10 \omega_\nu$ for the first vibrational transition. For the sake of comparison, we mention that single-molecule electronic strong coupling has been achieved with mode volumes of ~ 40 nm^3 [17], corresponding to $\lambda \approx 0.007$ a.u., and there are indications that effective sub-nm^3 mode volumes could be reached due to single-atom hot spots [18, 19], which would allow the single-mode coupling strength to reach values up to $\lambda \approx 0.05$ a.u..

In Fig. 6.2 we show the resulting rates with different values of the coupling constant in an Arrhenius plot, i.e., the logarithm of the rate divided by the temperature

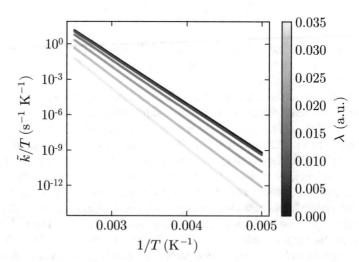

Fig. 6.2 Arrhenius plot for the rate dependence with temperature in the hybrid system for several light-matter coupling values. See main text for details

as a function of the inverse temperature. The straight lines in Fig. 6.2 confirm that the hybrid light–matter system follows the behavior described by the Eyring equation [15]

$$k = \kappa 2\pi k_B T e^{-\frac{E_b}{k_B T}}, \tag{6.12}$$

which connects the rate of a chemical reaction with the energy barrier E_b that separates reactants from products. This expression is Eq. (2.41) expressed in atomic units. Here, κ is a transmission coefficient, typically considered equal to one if nonadiabatic effects can be neglected close to the transition state.

We thus observe that even under vibrational strong coupling and the accompanying formation of vibro-polaritons, i.e., hybrid light–matter excitations, the reaction rate can still be described by an effective potential energy barrier. However, the effective height of the energy barrier is modified through the CQED effect of strong coupling, leading (for the studied model) to significantly reduced reaction rates. Although we treat here a single-mode and single-molecule system, these general observations agree with experimental studies [1, 2, 4].

6.3.2 CBOA-Based Model

We develop here a theory not based on full quantum rate calculations (which require the calculation of nuclear dynamics in $3N - 6$ dimensions) and that allows to make predictions beyond simple model systems. In the following we show that this can be achieved by applying (classical) transition state theory (TST) to the combined photonic–nuclear potential energy surfaces provided by the cavity Born–Oppenheimer approximation introduced in Sect. 2.3 and that we can get general results by combining CBOA and perturbation theory.

6.3.2.1 Cavity Born–Oppenheimer Surfaces

We now apply the cavity Born–Oppenheimer approximation to our system with the goal to get the cavity-PES of the ground state $\tilde{V}_g(R, q)$ in which the modifications when increasing λ are visible. We achieve this by diagonalizing the new electronic Hamiltonian $\hat{H}_e(\hat{x}; R, q) = \hat{H} - \frac{\hat{p}^2}{2} - \frac{\hat{p}^2}{2M}$. Conceptually, the photonic displacement corresponds to a single additional nuclear-like DoF. This allows to apply standard tools such as TST to obtain an estimation for reaction rates. With this theory it is only necessary to calculate the effective energy barrier for the reaction within the ground-state CBOA surface.

We use this for the Shin–Metiu model coupled to a cavity mode on resonance with the first vibrational transition. The two-dimensional PES $\tilde{V}_g(R, q)$ is shown in Fig. 6.3a for a coupling strength of $\lambda = 0.02$ a.u., which corresponds to a vibrational Rabi splitting of $\Omega_R \approx 0.05\omega_\nu$. The second panel, Fig. 6.3b, shows the minimum

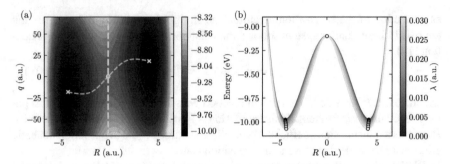

Fig. 6.3 a Two-dimensional ground-state PES in the cavity Born–Oppenheimer approximation for the Shin–Metiu model for $\lambda = 0.02$ a.u. and $\omega_c = 72.6$ meV. At $R = 0$ we show the dividing surface used to compute the reaction flux from reactant to product states. The gray dashed line curve corresponds to the energy path along $q_m(R)$, i.e., the minimum in q. **b** Value of the energy path $\tilde{V}_g(\mathbf{R}, q_m)$ for different values of the Rabi frequency, which is related to the coupling strength through $\Omega_R = \lambda \mu'_g(R_0)$, where the dipole derivative is evaluated at the minimum

Fig. 6.4 Energy barrier and rates ratio versus coupling strength for the case of a CBOA calculation (full lines) and for the effective energy barrier fitted from exact quantum rate calculations (dashed lines)

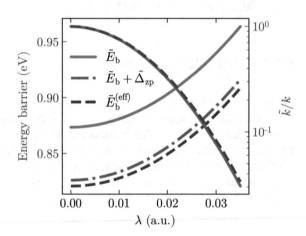

along q of this surface as a function of R, i.e., along the path indicated by the curved dashed line in Fig. 6.3a, for a set of coupling strengths λ that induce a Rabi splitting of up to $\Omega_R = 0.1\omega_\nu$. This path closely corresponds to the minimum energy path of the proton transfer reaction within the CBOA. As the coupling is increased, the minima become deeper, while the transition state (TS) at $R = 0$ stays unaffected. This leads to an effective increase of the reaction barrier $\tilde{E}_b = \tilde{V}_g(R_{TS}, q_{TS}) - \tilde{V}_g(R_{min}, q_{min})$, as shown in Fig. 6.4.

In this figure we also show the corresponding change in the rate predicted by Eq. (6.12). The full lines correspond to the energy barrier calculated within the CBOA (blue) and the corresponding rate (red) according to TST, while the dashed lines show the effective energy barrier $E_b^{(eff)}$ extracted from the fit to the Arrhenius plot Fig. 6.2 and the corresponding change in the rate obtained from the full quantum rate calculation above. As can be seen, the effective and CBOA energy barriers agree very well, with just an approximately constant overestimation of the barrier in CBOA due to quantum effects such as zero-point energy and tunneling, which remain unaffected by the cavity. This leads to excellent agreement for the change of the reaction rate obtained from the full quantum calculation and the CBOA-TST prediction. As expected from our previous discussion, the reaction rate of the hybrid cavity–molecule system decreases dramatically as the coupling increases due to the increase of the energy barrier height. Finally, we also calculate the CBOA energy barrier corrected by $\tilde{\Delta}_{zp}$, the difference between the zero-point vibrational frequencies at the minimum and transition states as obtained from the Hessian of the PES (disregarding the direction of negative curvature at the TS). This is shown as a dash-dotted line in Fig. 6.4, and considerably improves the absolute agreement with the effective barrier extracted from the full quantum rate calculations.

While we have up to now worked within a single-mode model, the CBOA actually makes it straightforward to treat multiple photonic modes. The ground state PES then parametrically depends on multiple parameters q_k, one for each mode, just as a realistic molecule depends on multiple nuclear positions $\mathbf{R}_i$. Similarly, the adiabatic surfaces are not harder to calculate than for the single-mode case, and minimization strategies can rely on the same approaches used in "traditional" quantum chemistry. We note that for a general cavity, the mode parameters can be obtained either by explicitly quantizing the modes (which is in general a difficult proposition) or, alternatively, by rewriting the spectral density of the light-matter coupling (proportional to the EM Green's function) as a sum of Lorentzians [28–31].

6.3.2.2 Perturbation Theory

As we have seen, the cavity Born–Oppenheimer approximation provides a convenient framework to evaluate cavity-induced changes in chemical reactivity based on energy barriers in electronic PES that are parametric in nuclear and photonic coordinates. In particular, the interaction term $\omega_c q \boldsymbol{\lambda} \cdot \hat{\boldsymbol{\mu}}$, with q as a parameter, is equivalent to that obtained from applying a constant external electric field. The cavity PES for arbitrary molecules can thus be calculated with standard quantum chemistry codes. However, obtaining the barrier in general still requires minimization of the molecular PES along the additional photon coordinate q (or coordinates q_k, if multiple modes are treated). If the coupling is not too large and the relevant values of q are small enough, the ground-state cavity PES can instead be obtained within perturbation theory, which up to second order in λ is given by

$$\tilde{V}_g(\mathbf{R}, q) \approx V_g(\mathbf{R}) + \frac{\omega_c^2}{2}q^2 + \lambda\omega_c q \mu_g(\mathbf{R}) - \frac{\lambda^2}{2}\omega_c^2 q^2 \alpha_g(\mathbf{R}), \qquad (6.13)$$

where $V_g(\mathbf{R})$ and $\mu_g(\mathbf{R})$ are the bare-molecule ground-state PES and dipole moment, respectively, while $\alpha_g(\mathbf{R})$ is the ground-state static polarizability (see Sect. 2.2),

$$\alpha_g(\mathbf{R}; \omega = 0) = 2 \sum_{m \neq g} \frac{|\mu_{m,g}(\mathbf{R})|^2}{V_m(\mathbf{R}) - V_g(\mathbf{R})}, \tag{6.14}$$

and encodes the effect of excited electronic levels, with $\mu_{m,g}(\mathbf{R})$ the transition dipole moment between bare-molecule electronic levels m and g. Obtaining the full ground-state cavity PES within this approximation then just requires the calculation of the bare-molecule ground-state properties $V_g(\mathbf{R})$, $\mu_g(\mathbf{R})$, and $\alpha_g(\mathbf{R})$.

In addition to providing an explicit expression for the CBO ground-state PES in terms of bare-molecule ground-state properties, the simple analytical dependence on q in Eq. (6.13) allows to go one step further and obtain explicit expressions for the local minima and saddle points (i.e., transition states). In these configurations, the following conditions are satisfied:

$$\partial_q \tilde{V}_g(\mathbf{R}, q) = 0, \tag{6.15a}$$

$$\partial_{\mathbf{R}} \tilde{V}_g(\mathbf{R}, q) = 0. \tag{6.15b}$$

These conditions yield a set of coupled equations that can be solved in order to find the configuration of the new critical points along the reaction path. The expression of Eq. (6.15a) gives the explicit condition

$$q_m(\mathbf{R}) = -\frac{\lambda}{\omega_c} \frac{\mu_g(\mathbf{R})}{1 - \lambda^2 \alpha_g(\mathbf{R})}, \tag{6.16}$$

which can be used to obtain the potential profile along the minimum in q,

$$\tilde{V}_g(\mathbf{R}, q_m) = V_g(\mathbf{R}) - \frac{\lambda^2}{2} \mu_g^2(\mathbf{R}) + \mathcal{O}(\lambda^4), \tag{6.17}$$

where we have dropped terms of order λ^4 since the perturbation-theory PES Eq. (6.13) is only accurate to second order. This shows that the energy barrier on the cavity PES (within second-order perturbation theory) can be calculated directly from the bare-molecule potential and permanent dipole moment. In Fig. 6.5, we analyze the validity of Eq. (6.17) for computing the barrier height within the Shin–Metiu model. It can be observed that perturbation theory works quite well for the whole range of couplings, with a relative error in the cavity-induced change of the energy barrier of about 10% for the largest considered couplings. Due to the exponential dependence of the rates on barrier height, this corresponds to an appreciable error in the rate constant, but still provides a reasonable estimate. Note that in the case of the Shin–Metiu model, the error of the energy barrier stems entirely from the change at the minimum configuration, as the transition state has zero dipole moment due to symmetry and is not affected by the cavity.

It is interesting to point out that Eq. (6.17) closely resembles the expression obtained in electric field catalysis where an external voltage is applied [32], or to electrostatic shifts provided by some catalysts [33]. This strategy exploits the Stark

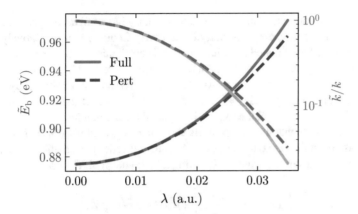

Fig. 6.5 Cavity Born–Oppenheimer energy barrier (purple) and relative change of reaction rates (yellow) for the Shin–Metiu model inside a cavity, calculated to all orders in the light-matter coupling strength λ (solid lines), and up to second order in perturbation theory (dashed lines)

effect, i.e., the energy shift observed in the presence of a static electric field, to induce changes in the energies of the transition state relative to the minimum configuration. As noted before, the CBOA corresponds to treating the influence of the cavity through an adiabatic parameter q determining the electric field strength. However, instead of being externally imposed, in our case the effective field, determined by Eq. (6.16), is the one induced in the cavity by the permanent dipole moment of the molecule itself. This also lends itself to an electrostatic interpretation of the effect.

In addition to the minimum energy barrier of the cavity PES itself, the effective energy barrier is also affected by the zero-point energy due to the quantization of nuclear and photonic motion (see Fig. 6.3). We can obtain its cavity-induced shift within perturbation theory by using Eq. (6.16) to rewrite Eq. (6.13) as

$$\tilde{V}_g(\mathbf{R}, q) = \tilde{V}_g(\mathbf{R}, q_m) + \frac{\omega_{\text{eff}}^2(\mathbf{R})}{2}(q - q_m(\mathbf{R}))^2, \tag{6.18}$$

where $\omega_{\text{eff}}(\mathbf{R}) = \omega_c - \frac{\lambda^2}{2}\omega_c\alpha_g(\mathbf{R}) + \mathcal{O}(\lambda^4)$, such that the photonic zero-point energy $\omega_{\text{eff}}(\mathbf{R})/2$ is decreased due to the polarizability of the molecule. We note that this only accounts for the quantization of the photonic motion along q. Indeed, there is an additional correction due to the vibrational contribution to the molecular polarizability. This can be obtained by diagonalizing the Hessian of Eq. (6.9) and calculating the zero-point energy, which up to second order is given by $-\frac{\omega_c\Omega_R^2}{4\omega_v(\omega_c+\omega_v)}$, where $\Omega_R = \frac{\lambda}{\sqrt{M}}\mu_g'(R_0)$ is the on-resonance vibrational Rabi splitting as discussed in above. As can be appreciated from Fig. 6.4, the contributions due to zero-point (photonic and vibrational) fluctuations only contribute negligibly to the change in reaction rate in the Shin–Metiu model.

In general, a significant change of polarizability (either electronic or vibrational, which can be comparable in some molecules [34–36]) from the equilibrium to the transition state configuration could lead to similarly large effects as a change in the permanent dipole moment, especially if the cavity frequency ω_c is relatively large. However, it can be estimated that the vibrational contribution to the zero-point energy shift is negligible for conditions typical for vibrational strong coupling. To be precise, at resonance $\omega_c = \omega_v$, this reduces to $-\Omega_R^2/(8\omega_v)$. Even for a relatively large vibro-polariton Rabi splitting of $\Omega_R \approx 0.2\omega_v$ [23, 37, 38], this contribution is of the order of $\approx 10^{-2}\omega_v$, and thus small compared to typical barrier heights.

Finally, we note that the energy shifts above can be straightforwardly generalized to the case of multiple cavity modes within second-order perturbation theory. This simply leads to a sum over modes k, giving a final energy shift

$$\delta E(\mathbf{R}) = -\sum_k \frac{\lambda_k^2}{2}\left(\mu_g^2(\mathbf{R}) + \frac{\omega_k}{2}\alpha_g(\mathbf{R})\right). \tag{6.19}$$

This general expression, which is just the second-order energy correction due to coupling to a set of cavity modes within the CBOA, corresponds to the well-known Casimir–Polder energy shift [39]. The additional cavity Born–Oppenheimer approximation, in which nonadiabatic transitions between electronic surfaces are neglected, amounts to the approximation that the relevant cavity frequencies ω_k are much smaller than the electronic excitation energies $V_m(\mathbf{R}) - V_g(\mathbf{R})$, such that only the (electronic) zero-frequency polarizability $\alpha_g(\mathbf{R})$ appears in the second term. In contrast, the first term depends only on the ground-state molecular permanent dipole moment $\mu_g = \langle g|\hat{\mu}|g\rangle$, which does not involve electronically excited states, and the CBOA thus does not amount to an additional approximation.

Interestingly, for cavities with a dipole-like field, the perturbative energy shifts obtained here correspond exactly to van der Waals forces [40]. We can easily demonstrate this for a general nanoparticle with a series of (bosonic) dipole resonances characterized by (vectorial) transition dipoles $\boldsymbol{\mu}_k$ and frequencies ω_k. In this case, the coupling operators $\boldsymbol{\lambda}_k$ at the molecular position $\mathbf{r}_m$ are determined by the static dipole–dipole interaction,

$$\boldsymbol{\lambda}_k = \sqrt{\frac{2}{\omega_k}}\left(\frac{3(\boldsymbol{\mu}_k \cdot \mathbf{r}_m)\mathbf{r}_m}{r_m^5} - \frac{\boldsymbol{\mu}_k}{r_m^3}\right). \tag{6.20}$$

For simplicity, we assume $\mathbf{r}_m$ to be along the x-axis, and all dipoles to be oriented along z, which leads to $\lambda_k = \sqrt{\frac{2}{\omega_k}}\frac{\mu_k}{r_m^3}$. By inserting this in Eq. (6.19) and using the definition of the zero-frequency polarizability of the nanoparticle $\alpha_n(0) = \sum_k \frac{2\mu_k^2}{\omega_k}$, we get that

$$\delta E(\mathbf{R}) = -\frac{\alpha_q(0)\mu_g^2(\mathbf{R})}{2r_m^6} - \sum_k \frac{\mu_k^2 \alpha_g(\mathbf{R})}{2r_m^6}. \tag{6.21}$$

where the first term corresponds exactly to the static energy of a permanent dipole μ_g at $\mathbf{r}_m$ with the induced dipole of a polarizable sphere at the origin (Debye force), and the second term corresponds to the London force [41].

Equation (6.19) is general for any kind of molecular process as long as the light–matter coupling is not too large. It demonstrates that the most relevant bare-molecule properties determining cavity-induced chemical reactions in the ground state are the permanent dipole moment and polarizability close to equilibrium, $\mu_g(\mathbf{R}_0)$ and $\alpha_g(\mathbf{R}_0)$, and transition state, $\mu_g(\mathbf{R}_{TS})$ and $\alpha_g(\mathbf{R}_{TS})$, configurations, and *not* the transition dipole moment of the vibrational excitation close to equilibrium, $\mu_\nu \propto \mu'_g(\mathbf{R}_0)$, that determines the Rabi splitting. In addition to changing reaction barriers, it should be noted that the cavity-induced modification could potentially lead to a plethora of diverse chemical modifications, such as a change of the relative energy of different (meta-)stable ground-state configurations and thus a change of the most stable configuration, or even the creation or disappearance of stable configurations. Furthermore, depending on the particular properties of the molecule, the cavity-induced change in the energy barriers can either lead to suppression or acceleration of chemical reactions.

6.3.3 Resonance Effects

The results presented above predict a change in the ground-state reactivity that is actually independent of the cavity photon frequency and in particular does not rely on any resonance effects between the cavity mode and the vibrational transitions of the molecule. Although the cavity PES can and does represent vibro-polariton formation through normal-mode hybridization, as discussed above and in Sect. 2.3, the subsequent TST used to predict changes in chemical reaction rates is an inherently classical theory and does not depend on the quantized frequencies of motion on the PES, and, as mentioned above, neither on the transition dipole moment between vibrational levels (determined by the derivative of the permanent dipole moment). While we have shown that TST agrees almost perfectly with full quantum rate calculations, where nuclear and photonic motion is quantized and polariton formation is thus included, all calculations above have been performed for the resonant case $\omega_c = \omega_\nu$.

We thus investigate whether there is any resonance effect on chemical ground-state reactivity by performing full quantum rate calculations for a wide range of cavity frequencies within the Shin–Metiu model. In Fig. 6.6, we represent the change $\tilde{k}/k$ in the calculated reaction rate of the coupled system relative to the uncoupled molecule as a function of ω_c, for three different coupling strengths λ. Here, the values at $\omega_c = \omega_\nu$ correspond to the results shown in Figs. 6.3 and 6.4. We observe that the cavity rates are essentially constant with the frequency, with only a small modulation ($\tilde{k}(\omega_c \to \infty) - \tilde{k}(\omega_c \to 0) \neq 0$) that becomes more important for larger couplings. For the cases represented in Fig. 6.6, this goes from a relative modulation of 0.4% for $\lambda = 0.005$ a.u. to a 7% modulation for $\lambda = 0.02$ a.u.. However, no resonance effects

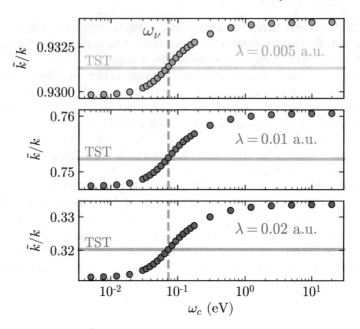

Fig. 6.6 Ratio between on- ($\tilde{k}$) and off-cavity (k) rates versus the cavity frequency for three different values of the coupling strength. We increase the density of points close to the vibrational frequency of the molecule $\omega_\nu \approx 72.6$ meV in order to explore potential resonance effects

are revealed close to the vibrational frequency of the molecule, ω_ν. At the same time, the vibrational frequency appears to be the relevant energy that separates the high- and low-frequency limits for the rates, with TST working particularly well exactly around that value. In the following, we show that both limits can be understood by different additional adiabatic approximations.

In the high-frequency limit, $\omega_c \gg \omega_\nu$, the photonic degree of freedom is fast compared to the vibrational one, and can thus be assumed to instantaneously adapt to the current nuclear position R. This implies that the photonic DoF can be adiabatically separated (just like the electronic ones), and nuclear motion takes place along an effective 1D-surface determined by the local minimum in q, i.e., along the path sketched in Fig. 6.3a, or, within lowest-order perturbation theory, along the surface defined by Eq. (6.17). Quantum rate calculations along this effective 1D PES indeed reproduce the reaction rate in the high-frequency limit perfectly. Furthermore, we note that in this limit, it becomes convenient to directly group the photonic and electronic degrees of freedom to obtain PoPES when performing the Born–Oppenheimer approximation. In particular, this approach leads to exactly the same expression for the effective ground-state PES.

In the low-frequency limit, $\omega_c \ll \omega_\nu$, on the other hand, the photonic motion is much slower than the vibrations and can also be adiabatically separated. The photons are now too slow to adjust their configuration and q can be assumed to stay constant during the reaction. The full quantum rate can then be obtained by performing a thermal average of independent 1D quantum rate calculations for each cut in q of the two-dimensional surface $\tilde{V}_g(R, q)$. Here, the (normalized) thermal weight at each q, $\mathcal{P}(q) = \exp(-\langle E \rangle(q)/k_B T)$, is calculated by calculating the average thermal energy of the system $\langle E \rangle(q)$ for constant q. Again, this approximation agrees perfectly with the full quantum rate calculation for $\omega_c \to 0$.

These results imply that, on the single-molecule level, the formation of vibro-polaritons when $\omega_c \approx \omega_\nu$ is not actually required or even relevant for the cavity-induced change in ground-state chemical structure and reactivity. This fact can be appreciated by a simple intuitive argument: vibrational strong coupling primarily occurs with the lowest vibrational transitions close to the equilibrium configuration, while chemical reactions that have to pass an appreciable barrier are typically determined by the properties of the involved transition state, and the associated barrier height relative to the ground-state configuration. In general, neither of these are related to the properties of the lowest vibrational transitions (i.e., curvature of the PES and derivative of the dipole moment at the minimum).

The absence of resonance effects can also be appreciated through the connection to the well-known material-body-induced potentials obtained within perturbation theory. For example, as we have demonstrated above, if the EM mode is well-approximated by a point-dipole mode, the obtained energy shift in the cavity PES can be rewritten as a van-der-Waals-like interaction between the permanent dipole moment of the molecule and the dipole it induces in the nanoparticle. This corresponds to the Debye force. In turn, the zero-point energy of the EM field reproduces the London dispersive force due to vacuum fluctuations, and depends on the polarizability of the molecule. For an arbitrary EM environment, this effect can also be directly linked to Casimir–Polder forces [39, 42], which exactly correspond to the generalization of emitter–emitter interactions to arbitrary material bodies (e.g., cavities). In particular, within the perturbative regime, the applicability of Casimir–Polder approaches could also be used to replace the explicit sum over modes k by integrals involving the EM Green's function [43, 44], which is readily available for arbitrary structures. This provides an additional argument for the absence of resonance effects in our calculations, as (ground-state) Casimir–Polder forces are well-known not to depend on resonances between light and matter degrees of freedom.

While we do not explicitly treat the situation in recent experiments on the modification of ground-state reactions by vibrational strong coupling (which were found to depend strongly on resonance conditions [1–4]), we believe that our results indicate that the resonance-dependent effects cannot be explained by a straightforward modification of ground-state reaction energy barriers at thermal equilibrium, as these would be captured by TST within the CBOA also in a many-mode, many-molecule setting.

6.4   Modifying Chemistry in Realistic Systems

Up to now we have studied ground-state chemistry in a general way, obtaining useful expressions such as Eq. (6.19) which describe the Casimir–Polder energy shift in arbitrary cavities. In order to quantitatively analyze the results, we used the Shin–Metiu model coupled to a single-mode cavity. Indeed, this is a great simplification of the system, so we here demonstrate the power of this theory by applying it to more realistic examples. In this section we analyze two different scenarios. First, we study a multi-mode cavity that is experimentally available and can reach large enough values of the coupling strength so that relevant chemical changes are visible. Then, we will apply the theory to a realistic scenario of internal rotation in the 1,2-dichloroethane molecule, showing not only that this theory can be combined with quantum chemistry approaches in order to predict more complex and relevant chemical changes, but also that reactions can be both inhibited or catalyzed depending on the molecular properties.

6.4.1   Multi-mode Cavity: Nanoparticle on Mirror

To demonstrate that the effects predicted above can be significant in realistic systems, we treat a nanoparticle-on-mirror cavity with parameters taken from the experiment in [17]. This consists of a spherical metallic nanoparticle (radius $R = 20$ nm) separated by a small gap from a metallic plane, see the inset of Fig. 6.7. In this system, there is a series of multipole modes coupled to the molecule [31], with nontrivial behavior. Although several strategies can be employed to obtain the quantized light modes in this system [19, 31], we instead exploit that the dominant contribution we found above is due to Debye-like electrostatic forces induced by the permanent molecular dipole, and thus simply solve the electrostatic problem. To be precise, we calculate the energy shift of a permanent dipole in this cavity as obtained by its interaction with the field it induces in the cavity itself. Due to the simple involved geometric shapes (a sphere and a plane), this can be achieved by the technique of image charges and dipoles.

In this nanocavity (see inset of Fig. 6.7) a permanent dipole will generate an infinite series of image dipoles in both the sphere and the plane due to successive "reflections" of each image dipole on both components of the cavity. In practice, this infinite converging series can be truncated after a finite number of terms to obtain any desired degree of accuracy. Considering both a charge q and a dipole μ at position $\mathbf{r}$ relative to the center of a perfectly conducting grounded sphere of radius R, the resulting images will be located at $\mathbf{r}' = (R/r)^2 \mathbf{r}$ (where $r = |\mathbf{r}|$) and consist of a charge and dipole given by

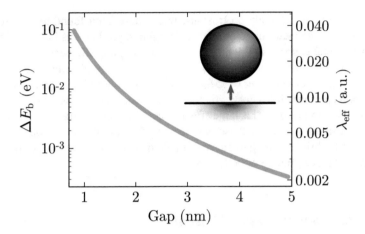

Fig. 6.7 Change of the energy barrier for the Shin-Metiu model inside a nanoparticle-on-mirror cavity as a function of gap size. The right y-axis shows the corresponding values of the effective single-molecule coupling strength $\lambda_{\text{eff}} = \sqrt{\sum_k \lambda_k^2}$. Inset: Illustration of nanoparticle-on-mirror cavity geometry, with a single molecule placed in the nanogap between a planar metallic surface and a small metallic nanoparticle of radius $R = 20$ nm

$$q' = -\frac{R}{r}q + \frac{R}{r^3}\mathbf{r} \cdot \boldsymbol{\mu}, \tag{6.22a}$$

$$\boldsymbol{\mu}' = \left(\frac{R}{r}\right)^3 \left[\frac{2\mathbf{r}\,(\mathbf{r} \cdot \boldsymbol{\mu})}{r^2} - \boldsymbol{\mu}\right]. \tag{6.22b}$$

Here, it is important to take into account that the image of a dipole in a sphere always consists of both a charge and a dipole. The corresponding expressions for a plane can be obtained by simply taking $R \to \infty$ (and moving the center of the sphere accordingly to keep the planar surface fixed). The cavity-induced energy shift of the dipole is then given by $U = -\frac{1}{2}\mathbf{E}_{\text{ind}} \cdot \boldsymbol{\mu}$, where $\mathbf{E}_{\text{ind}}$ is the total field generated by all image dipoles and charges, and the factor $\frac{1}{2}$ is due to them being induced.

We now rely again on perturbation theory, i.e., we assume that the molecular rearrangement due to its self-induced field is negligible. Within this approximation, the energy shift we obtain from the purely electrostatic calculation is equivalent to the term proportional to μ_g^2 in Eq. (6.19). The corresponding change ΔE_b in the height of the energy barrier for the Shin–Metiu molecule is shown in Fig. 6.7 as a function of the gap size (as a point of reference, the estimated gap size in [17] is 0.9 nm). We find that the change in energy barrier can be significant, corresponding to a change of the reaction rate by an order of magnitude or more (cf. Fig. 6.5). For comparison, in the figure we also show the effective coupling strength $\lambda_{\text{eff}} = \sqrt{\sum_k \lambda_k^2}$ corresponding to each gap size. This value corresponds to the coupling strength in a single-mode cavity that would give the same total energy shift as obtained in this realistic multi-mode cavity. We note that we have here treated a perfect spherical nanoparticle, and did not include atomic-scale protrusions, which have been found to lead to even larger field confinement due to atomic-scale lightning rod effects [18, 19, 45]. For the experimental gap size of 0.9 nm, the effective coupling still becomes as large as $\lambda_{eff} \approx 0.031$ a.u., corresponding to $V_{eff} = 4\pi/\lambda_{eff}^2 \approx 1.9$ nm^3. This corresponds to a change in the energy barrier of $\Delta E_b \approx 0.07$ eV for the Shin-Metiu model within second-order perturbation theory, which starts to break down at these couplings, as we previously saw in Fig. 6.5. This large effective coupling demonstrates the importance of the multi-mode nature of these cavities and the contribution of optically dark modes, as the "bright" nanogap plasmon mode that is seen in scattering spectra has an estimated mode volume of ≈ 40 nm^3.

6.4.2   1,2-Dichloroethane Molecule

We now apply the CBOA-TST theory to treat the internal rotation of 1,2-dichloroethane. In order to obtain the ground-state cavity PES under strong light–matter coupling, we calculate the (ground and excited-state) bare-molecule potential energy surfaces and permanent and transition dipole moments for a scan along the rotation angle (defined as the Cl-C-C-Cl dihedral angle). For simplicity, we here use the relaxed ground-state configuration of the bare molecule for each rotation angle, i.e., we neglect cavity-induced changes in DoF different from the internal rotation angle. The molecular properties are obtained with density functional theory calculations with the B3LYP [46] hybrid exchange-correlation functional and the 6-31+G(d) basis set. Excited states were computed with time-dependent density functional theory within the Tamm–Dancoff approximation [47]. All calculations were performed with the TeraChem package [48, 49].

The rather simple 1,2-dichloroethane molecule presents several characteristic configurations along the rotation of the chlorine atoms around the axis defined by the carbon-carbon bond (see top of Fig. 6.8). It thus constitutes an excellent model system to show several possible effects induced by coupling to a cavity. In Fig. 6.8a we present the calculated ground state energy landscape and dipole moment, while some relevant configurations are shown at the top. Analogously to the Shin–Metiu case, we present the path of minimum energy along q in Fig. 6.8b, but here calculated within perturbation theory, Eq. (6.17). We have explicitly checked that the contribution due to London forces is negligible here as well, and focus on the Debye-like contribution in the following. We see that the most stable configuration ($\theta = 180°$) shows no change due to the absence of a permanent dipole moment, while the most unstable one presents a large energy shift. Therefore the different energy barriers of the system, represented versus the coupling strength in Fig. 6.8c, are altered significantly. Here we compare the energy barriers as predicted by perturbation theory (dashed lines) with the ones from a full diagonalization of the electronic Hamiltonian within the CBOA (full lines). In order to perform a full calculation we have calculated the electronic potential energy surfaces and the full dipole moment operator for a basis of 17 electronic states. We also indicate the points at which the coupling leads to

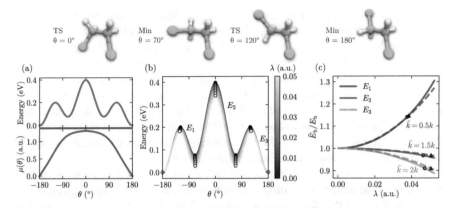

Fig. 6.8 Top: Different configurations along the internal rotation of 1,2-dichloroethane. **a** Energy landscape and dipole moment of the molecule. **b** Modified energy path for minimum q for different coupling strengths. The energy barriers of the bare molecule are defined as $E_1 = V(120°) - V(70°)$, $E_2 = V(0°) - V(70°)$, and $E_3 = V(120°) - V(180°)$. **c** Relative modification of the energy barriers depending on the coupling strength for the full calculation (full lines, circles) and for perturbation theory (dashed lines, triangles). The marked points indicate relevant changes in the rate

important changes in the relative rates calculated with TST, i.e., the coupling/energy at which we achieve either suppression of $\tilde{k}/k = 0.5$ or enhancement of $\tilde{k}/k = 1.5$ or 2. We see that in the case of perturbation theory (triangles) the energy changes are slightly underestimated and thus larger couplings are needed to reach the same rate change as in the full calculation (circles).

As can be clearly seen, this still relatively simple molecule shows several different kinds of phenomena. We see that the reaction rate out of the global minimum at $\theta = 180°$, corresponding to E_3, is increased. On the other hand, E_1 increases and the local minimum situated at $\theta = 70°$ is thus stabilized. Figure 6.8b suggests that this effect could potentially become more dramatic for larger couplings than treated here, as $\theta = 70°$ could become the new global minimum of the system. Finally, it is worth noting that the locations of the minima in energy also change for larger couplings. This shift is most noticeable for the minimum at $\theta = 70°$, which transforms to $\tilde{\theta} \approx 68°$ for $\lambda = 0.05$ a.u..

6.5 Collective Effects

We now turn to the description of collective effects, i.e., the case of multiple molecules. For simplicity, we again restrict the discussion to a single cavity EM mode. As discussed above, the single-molecule effects we have discussed up to now only become significant for coupling strengths $\lambda = \sqrt{4\pi/V_{\text{eff}}}$ corresponding to the smallest available plasmonic cavities, which typically operate at optical frequencies.

However, typical experimental realizations of vibrational strong coupling are performed in micrometer-size cavities filled with a large number of molecules [1, 23, 37, 50]. In this case, the per-molecule coupling λ is so small that the single-molecule effects discussed above are completely negligible. For strong coupling and the associated formation of vibro-polaritons, the coherent response of all molecules leads to a collective enhancement of the Rabi splitting $\Omega_{R,col} = \sqrt{N}\Omega_R$. However, as we have seen that the cavity-induced modification of the single-molecule ground state does not depend on the formation of polaritons, it is not a priori obvious whether this collective enhancement of the Rabi splitting also translates to cavity-induced collective modifications of the effective reaction barrier.

We thus repeat the analysis performed for the single-molecule case above for the case of multiple molecules, working directly within the cavity Born–Oppenheimer approach. We note that the arguments for its applicability for treating ground-state chemical reactions translate straightforwardly from the single- to the many-molecule case. For N identical molecules, the CBO light–matter interaction Hamiltonian becomes

$$\hat{H}_e^{(N)} = \frac{\omega_c^2}{2}q^2 + \sum_i \left(\hat{H}_e(\hat{\mathbf{x}}_i; \mathbf{R}_i) + \omega_c q \boldsymbol{\lambda}_i \cdot \hat{\boldsymbol{\mu}}(\hat{\mathbf{x}}_i; \mathbf{R}_i) \right) + \sum_{i,j} \hat{H}_{dd}(\hat{\mathbf{x}}_i, \hat{\mathbf{x}}_j; \mathbf{R}_i, \mathbf{R}_j),$$

(6.23)

where $\hat{H}_{dd}$ accounts for direct intermolecular (dipole–dipole) interactions. We stress that we again assume that only a single cavity mode is significantly coupled to the molecules. The cavity-mediated dipole–dipole interaction is thus fully contained within the light–matter coupling term, and $\hat{H}_{dd}$ corresponds to the free-space expression [51]. In the following discussion, we will again use lowest-order perturbation theory to obtain analytical insight. The cavity–molecule and dipole–dipole interaction terms are then independent additive corrections. We first focus on the cavity-induced effects, and will discuss the influence of direct dipole–dipole interactions later, in particular when studying a prototype implementation: A nanosphere surrounded by a collection of molecules. For simplicity of notation, we again use scalar quantities to indicate the component of the dipole along the field direction, but keep the index ϵ to make this explicit, i.e., $\boldsymbol{\lambda}_i = \lambda_i \epsilon_i$ and $\epsilon_i \cdot \boldsymbol{\mu}(\mathbf{R}_i) = \mu_\epsilon(\mathbf{R}_i)$, so that we can rewrite the interaction term of the Hamiltonian as $\omega_c q \sum_i^N \lambda_i \hat{\mu}_\epsilon(\hat{\mathbf{x}}_i; \mathbf{R}_i)$. The full Hamiltonian now corresponds to a many-body problem even for simple model molecules. Within second-order perturbation theory, the new (many-molecule) ground-state cavity PES is

$$\tilde{V}_g^{(N)}(\mathbf{R}_t, q) = \sum_i V_g(\mathbf{R}_i) + \frac{\omega_c^2}{2}q^2 + \omega_c q \sum_i \lambda_i \mu_{g,\epsilon}(\mathbf{R}_i) - \frac{\omega_c^2}{2}q^2 \sum_i \lambda_i^2 \alpha_{g,\epsilon\epsilon}(\mathbf{R}_i),$$

(6.24)

where $\mathbf{R}_t = (\mathbf{R}_1, \mathbf{R}_2, \ldots, \mathbf{R}_N)$ collects the nuclear configurations of all the molecules. With this result, we can again apply the usual conditions for finding critical points in order to analytically find the minimum along q and the corresponding total energy of the hybrid system up to second order in λ_i,

$$\tilde{V}_{\mathrm{g}}^{(N)}(\mathbf{R}_t, q_{\mathrm{m}}) = \sum_i V_{\mathrm{g}}(\mathbf{R}_i) - \frac{1}{2}\left(\sum_i \lambda_i \mu_{\mathrm{g},\epsilon}(\mathbf{R}_i)\right)^2. \tag{6.25}$$

It can be seen that the cavity-induced shift depends on the square of the sum of the (coupling-weighted) permanent dipole moments of the molecules, not on the sum of their squares. Assuming perfect alignment and identical configurations for all molecules, this gives an energy shift $-N^2\bar{\lambda}^2|\mu_{\mathrm{g}}(\mathbf{R})|^2$, where $\bar{\lambda} = \frac{1}{N}\sum_i \lambda_i$ is the average coupling. The per-molecule energy shift is then linear in N, indicating collective enhancement of the molecule-cavity interaction. In contrast, the London-force-like change in zero-point energy due to the modification of the effective cavity frequency is additive,

$$\omega_{\mathrm{eff}} = \omega_c - \frac{\omega_c}{2}\sum_i \lambda_i^2 \alpha_{\mathrm{g},\epsilon\epsilon}(\mathbf{R}_i) + \mathcal{O}(\lambda_i^4), \tag{6.26}$$

with a total zero-point energy shift $\frac{1}{2}(\omega_{\mathrm{eff}} - \omega_c)$ proportional to N, and shows no collective enhancement for single-molecule reactions. It is interesting to note that the connection between polarizability and the dielectric function of a material through the Clausius–Mossotti relation suggests that this energy shift is equivalent to the change of mode frequency due to the refractive index of the collection of molecules. The shift in cavity mode frequencies due to refractive index changes after chemical reactions is exactly the effect used in experiments to monitor reaction rates under vibrational strong coupling [1, 2, 4]. We also mention that at higher levels of perturbation theory, cavity-mediated contributions analogous to the Axilrod–Teller potential, i.e., van-der-Waals interactions between three emitters, appear in the intermolecular potential [44, 52].

Based on Eq. (6.25), we can analyze the effect of the cavity on the reaction rate of a single molecule within the ensemble. This is determined by the energy difference between minimum-energy and transition-state configurations of that molecule, with the other molecules fixed in a stable configuration (here chosen to be the minimum for all of them). For simplicity, we assume that the critical configurations $\mathbf{R}_{\mathrm{Min}}$ and $\mathbf{R}_{\mathrm{TS}}$ of the coupled system are equal to the uncoupled ones (as we have seen above, the shifts are generally small). We can then directly express the change in the energy barrier of the moving molecule (chosen to be molecule $i = 1$ here) as

$$\tilde{E}_{\mathrm{b}} = E_{\mathrm{b}} - \frac{\lambda_1^2}{2}\left(\mu_{\mathrm{g},\epsilon}^2(\mathbf{R}_{1,\mathrm{TS}}) - \mu_{\mathrm{g},\epsilon}^2(\mathbf{R}_{1,\mathrm{Min}})\right) -$$

$$\lambda_1\left(\sum_{i=2}^N \lambda_i \mu_{\mathrm{g},\epsilon}(\mathbf{R}_{i,\mathrm{Min}})\right)\left(\mu_{\mathrm{g},\epsilon}(\mathbf{R}_{1,\mathrm{TS}}) - \mu_{\mathrm{g},\epsilon}(\mathbf{R}_{1,\mathrm{Min}})\right). \tag{6.27}$$

This expression can be straightforwardly interpreted, with the first part corresponding to the Debye-like interaction of molecule 1 itself with the cavity, and the second part corresponding to the cavity-mediated interaction of molecule 1 with all other molecules (which itself can be understood as the sum of two equal contributions, the interaction of the moving molecule with the cavity field induced by all other molecules, as well as the interaction of all other molecules with the cavity field induced by the molecule). Within perturbation theory, this Debye-like energy shift is again equivalent to the electrostatic energy, in this case that of a collection of permanent dipoles interacting with the cavity, i.e., a material structure. This makes the connection to electric field catalysis [32] even more direct, with the difference that the electric field is not generated by applying an external voltage, but represents the cavity-enhanced field of all the other molecules. The fact that the main contribution is just the electrostatic energy shift also demonstrates the equivalence of our results to the approach of taking into account non-resonant effects through cavity-modified dipole–dipole and dipole-self interactions [51].

To treat the dependence on molecular orientations explicitly, we define the alignment angle θ_i for each molecule through $\mu_{g,\epsilon}(\mathbf{R}_i) = |\boldsymbol{\mu}_g(\mathbf{R}_i)| \cos \theta_i$. Inserting this in Eq. (6.27), we obtain

$$\tilde{E}_b = E_b - \frac{\lambda_1^2}{2} \left(\mu_{g,\epsilon}^2(\mathbf{R}_{1,\mathrm{TS}}) - \mu_{g,\epsilon}^2(\mathbf{R}_{1,\mathrm{Min}}) \right) -$$
$$N' \bar{\lambda}^2 \langle \cos \theta \rangle' |\boldsymbol{\mu}_g(\mathbf{R}_{\mathrm{Min}})| \lambda_{r,1} \left(\mu_{g,\epsilon}(\mathbf{R}_{1,\mathrm{TS}}) - \mu_{g,\epsilon}(\mathbf{R}_{1,\mathrm{Min}}) \right),$$
$$(6.28)$$

where $\lambda_{r,i} = \lambda_i / \bar{\lambda}$ is the relative coupling of molecule i, $\langle \cos \theta \rangle = \frac{1}{N} \sum_i \lambda_{r,i} \cos \theta_i$ is the coupling-weighted average orientation angle, and primed quantities indicate that only molecules 2 to N are taken into account (for $N \gg 1$, they can be replaced by unprimed quantities). We obtain a term proportional to the number of molecules N, i.e., there is a collective effect on the single-molecule energy barrier that is reminiscent of the collective Rabi splitting, $N\lambda^2 \propto \Omega_{R,\mathrm{col}}^2$. Note that the collective change of the energy barrier still depends on the molecule having a different permanent dipole moment in the transition and minimum configuration. Furthermore, it requires the molecules not participating in the reaction to have a non-zero permanent dipole moment and an average global alignment, such that $\langle \cos \theta \rangle \neq 0$. This could be achieved by fixing the molecular orientation by, e.g., growing self-assembled monolayers [53] or using DNA origami [54, 55], or for molecules that can be grown in a crystalline phase, such as anthracene [56] (although polar molecules tend not to grow into crystals with a global alignment [57]). Another strategy to achieve alignment under strong coupling that has been successfully used experimentally is to align molecular liquid crystals through an applied static field [58]. However, for general disordered media such as polymers or molecules flowing in liquid phase [1, 50], the angular distribution is typically isotropic, leading to $\langle \cos \theta \rangle \approx 0$. In that case, our theory predicts that no collective effect on reactivity should be observed unless the cavity itself induces molecular orientation (see below). We note for completeness that the collective Rabi splitting depends on the average of the squared z-component of the transition dipole moments, i.e., $\langle \cos^2 \theta \rangle$, which is nonzero unless all molecules

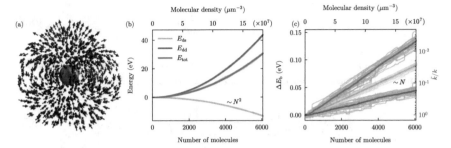

Fig. 6.9 **a** Sketch of the model system of a collection of molecules distributed around a metal nanosphere with a diameter of 8 nm. The molecules are placed randomly at distances from 1 nm to 16 nm to the surface of the sphere, with the (permanent) dipoles aligned along the direction of the field of the sphere's z-oriented dipole mode. **b** Energy due to the dipole-sphere (E_{ds}) and dipole-dipole (E_{dd}) interactions in the system within perturbation theory as a function of number of molecules N, as well as their sum (E_{tot}). **c** Change in energy barrier and corresponding change in reaction rate at room temperature for the most strongly coupled molecule, also resolved into contributions from dipole-sphere and dipole-dipole interactions. In both panels **b** and **c**, the slightly transparent lines correspond to different random realizations of the system, with the averages in solid lines

are aligned perpendicular to the electric field of the cavity mode, and equal to $1/3$ for isotropic molecules.

In order to test the strength of the collective effect in real cavity, and to compare it with the effect of direct (free-space) dipole–dipole interactions, we now treat a specific configuration, as depicted in Fig. 6.9a: A nanocavity represented by a metallic sphere of diameter $d = 8$ nm, surrounded by a collection of Shin–Metiu molecules, located at distances from 1 nm to 16 nm from the sphere. We place a collection of up to $N = 6000$ molecules at random positions within that volume, imposing a minimum distance of 1.5 nm between the molecules.

Let us first discuss the treatment of the spherical nanocavity. We describe the metal sphere[4] it using a Drude dielectric function, which allows us to approximate it as a three-mode cavity; the dipolar localized surface plasmon resonances aligned along x, y, and z [59, 60]. Higher order multipole modes only couple significantly to emitters that are very close to the surface [29, 61]. In this regime we can treat the nanosphere as a point dipole, in which the direction-independent polarizability of the sphere is given by [62]

$$\alpha_S(\omega) = a^3 \frac{\epsilon(\omega) - 1}{\epsilon(\omega) + 2},$$
(6.29)

where a is the radius of the sphere. We can then consider two general models for the dielectric function. The first is a metallic Drude model dielectric function without losses, $\epsilon_m(\omega) = 1 - \omega_p^2/\omega^2$. The polarizability of the sphere can then be rewritten as

---

[4]Note that the description used here can also represent a dielectric sphere with a single resonance, such as a phonon mode.

$$\alpha_S(\omega) = \frac{a^3 \omega_0^2}{\omega_0^2 - \omega^2}, \tag{6.30}$$

where $\omega_0 = \omega_p/\sqrt{3}$. This is identical to the polarizability of a single-mode quantum oscillator at frequency ω_0 with transition dipole moment $\mu_{eg} = \sqrt{\omega_0 a^3/2}$ [63],

$$\alpha_q(\omega) = \mu_{eg}^2 \left(\frac{1}{\omega_0 - \omega} + \frac{1}{\omega_0 + \omega} \right) = \frac{a^3 \omega_0^2}{\omega_0^2 - \omega^2}. \tag{6.31}$$

Here, spherical symmetry implies that there are three degenerate quantum oscillators, corresponding to the quantized localized surface plasmon resonances in this case, directed along three orthogonal axes (e.g., x, y, and z). The second possibility is that the dielectric function is instead given by Lorentzian function representing a material resonance (e.g., a phonon mode) at frequency ω_{ph} and with resonator strength characterized by ω_f, i.e., $\epsilon(\omega) = 1 + \frac{\omega_f^2}{\omega_{ph}^2 - \omega^2}$, we again get the same polarizability by using $\omega_0^2 = \omega_{ph}^2 + \frac{\omega_f^2}{3}$ and $\mu_{eg} = \omega_f \sqrt{\frac{a^3}{6\omega_0}}$, with the quantized mode now corresponding to a localized surface phonon polariton resonance.

We furthermore assume that all molecules are aligned perfectly with the electric field of the z-oriented dipolar mode of the sphere. In this configuration, the sum over x- and y-oriented fields at the origin cancels out for large N. For these directions, there is thus no Debye-like collective effect, and we can restrict our attention to just a single mode of the sphere (the z-oriented dipole mode). For the sake of completeness, we additionally check explicitly that solving the full electrostatic problem, i.e., including all modes of the sphere by using the method of image dipoles using the expressions in Eq. (6.22), gives very similar results to the ones presented here. As mentioned above, within perturbation theory, where the Debye-force-like contribution can be understood within a fully electrostatic picture, it is straightforward to include the direct (free-space) permanent-dipole–permanent-dipole interaction, as it is simply a further additive electrostatic contribution. In Fig. 6.9b, we show the total electrostatic energy of the system, as well as the relative contributions due to molecule–sphere and direct molecule–molecule interactions, as a function of N. For the configuration considered here, for which we have not performed any optimization of total energy, the dipole–dipole interactions give a positive contribution to the total energy that is significantly larger than the collective dipole–sphere interaction. The relative strength of dipole–dipole and dipole–sphere interactions depends on the details of the configuration, and we have checked that, e.g., it is also possible to maintain the same collective interaction while obtaining an overall negative contribution from dipole–dipole interactions by not choosing random positions as we did for simplicity.

In contrast to the total energy, the change in energy barrier predicted by Eq. (6.27) for the most strongly coupled molecule of the ensemble is dominated by the (collective) sphere–dipole interactions, as shown in Fig. 6.9c. The barrier height indeed increases approximately linearly with N, with changes of up to ≈ 0.09 eV due to the

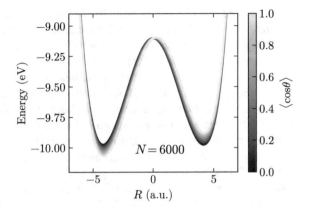

Fig. 6.10 Alignment dependence of the cavity Born–Oppenheimer PES along the photonic minimum path q_m for the molecule in Fig. 6.9c, with all other molecules fixed to the equilibrium position R_{min}

cavity-mediated interaction, and an associated suppression of the reaction rate by a factor of ≈ 30 at room temperature. In the geometry treated here, the energy shift of the target molecule due to dipole–dipole interactions with the other molecules also increases linearly with N, as the molecular dipoles combine to all act in the same direction at the sphere location, with an effect that is roughly half of the cavity-mediated interaction. As mentioned above, the details depend strongly on the configuration and cavity properties, and in particular, it is also possible to choose configurations where the direct dipole–dipole interactions dominate. We finally note that no simple configuration was found where the cavity-mediated interactions were significantly larger than direct dipole–dipole interactions.

While the barrier height increases here, the effect we predict can also lead to a decrease, for example in the case that the transition-state dipole moment is larger than in the minimum configuration, cf. Eq. (6.28). This would be expected, e.g., in dissociation reactions in which the molecule splits into two partially charged fragments, and is also seen for the back reaction from the right to left minimum in the Shin–Metiu model for the case that all other molecules are in the leftmost minimum (see Fig. 6.10).

For comparison, Fig. 6.10 shows the effect of average alignment for the sphere-molecule system considered above, for the case of $N = 6000$ molecules corresponding to a molecular density of $\approx 2 \cdot 10^8\ \mu m^{-3}$. It displays the CBO PES within second-order perturbation theory as a function of R_1, with all other molecules fixed in the minimum configuration, and along the photonic minimum $q = q_m$. For $\langle \cos \theta \rangle = 1$, this demonstrates that the collective cavity effect on the surface is significant, with the position of the critical points shifting compared to the bare molecule. For the Shin–Metiu model studied here, the barrier height is actually increased compared to the approximate prediction Eq. (6.27), which does not take into account these shifts. In contrast, when there is no average orientation, $\langle \cos \theta \rangle = 0$, the effect on the surface is minimal and is reduced to the single-molecule result.

The single-molecule energy shifts we predict for perfect alignment can be significant. This implies that the molecules, if they are free to rotate in place, could lower their energy by aligning with the electric field of the cavity mode, which could possibly lead to self-organization (for the example system above, this also requires breaking of the overall spherical symmetry). The details of this effect depend on the precise setup, such as the cavity material and shape, molecular and solvent properties, etc., and would require a more complete treatment taking thermodynamical effects and free energy into account [64, 65], which is beyond the scope of the current work. However, we mention that it has recently been shown that strong coupling and the associated formation of polaritons itself could lead to alignment due to the associated decrease of the lower polariton energy, provided that a significant fraction of molecules are excited to lower polariton states [66, 67]. Although thermal excitation can be efficient for vibrational strong coupling due to the relatively low energies of vibro-polaritons, on the order of a few times the thermal energy $k_B T$, it should be noted that the arguments in [66, 67] do not directly translate to thermal-equilibrium situations. In that case, a change in state energy due to improved orientation also leads to a change in population, with the average energy per degree of freedom staying constant and thus no net energy gain.

Finally, we mention that in contrast to the single-molecule case, the generalization of the above arguments to many cavity modes is not straightforward, and the results are thus not directly applicable to, e.g., Fabry–Perot cavities with a continuum of modes following a dispersion relation as a function of the in-plane wave vector, as employed in existing experiments [1–4]. Our results indicate that solving the electrostatic problem (where all modes are implicitly taken into account) should predict the changes in energy barriers, but, e.g., the scaling with number of molecules is not immediately obvious, and as mentioned above, statistical effects should be treated more carefully. Only for the special case that all modes have the same electric field distribution (e.g., different dipolar resonances of a small nanoparticle), the sum over modes can be performed straightforwardly.

6.6   Modifications of the Ground-State Structure

In this last section we discuss the influence of the cavity on the equilibrium configuration of the molecules. The theory developed in this chapter is very well-suited to study these kinds of changes. For the single-molecule case, it is straightforward to tackle this problem by looking at the condition for critical points in Eq. (6.15b), which states that the change in the equilibrium configuration is given by the equation $\partial_{\mathbf{R}} V_{\text{g}}(\mathbf{R}) - \lambda^2 \mu_{\text{g}}(\mathbf{R}) \partial_{\mathbf{R}} \mu_{\text{g}}(\mathbf{R}) = 0$. Although this depends on the specifics of the molecule, we can still estimate the change around the bare-molecule equilibrium position by doing a harmonic approximation for the ground-state PES, i.e., $V_{\text{g}}(R) \approx V_{\text{g}}(R_0) + \frac{1}{2} M \omega_\nu^2 (R - R_0)^2$, in which, for simplicity, we consider only one internal degree of freedom. For the dipole moment, we can do a first-order expansion, i.e., $\mu_{\text{g}}(R) \approx \mu_{\text{g}}(R_0) + \partial_R \mu_{\text{g}}(R_0)(R - R_0)$, which is equivalent to limit the molecule to single-phonon transitions [23]. Here, we consider only the projection of the dipole moment in the direction of the electric field $\mu_g(R) \equiv \mu_{g,\epsilon}(R) = \boldsymbol{\mu}_g(R) \cdot \boldsymbol{\epsilon}$. The new equilibrium configuration is

$$R_{0,\text{new}} \approx R_0 + \frac{\lambda^2}{M\omega_\nu^2}\mu_g(R_0)\partial_R\mu_g(R_0) + \mathcal{O}(\lambda^4). \qquad (6.32)$$

We see that the field generates a correction to the original minimum-energy configuration that depends both on the equilibrium dipole moment and on its first derivative, related to the vibrational transition dipole moment.

In the case for N molecules we get conditions similar to Eq. (6.15b) for every molecule, resulting in a set of coupled differential equations in which each single-molecule condition reads

$$\partial_{\mathbf{R}_i} V_g(\mathbf{R}_i) - \lambda_i^2 \mu_g(\mathbf{R}_i)\partial_{\mathbf{R}_i}\mu_g(\mathbf{R}_i) - \lambda_i \partial_{\mathbf{R}_i}\mu_g(\mathbf{R}_i)\left(\sum_{j\neq i}\lambda_j\mu_{ig}(\mathbf{R}_j)\right) = 0. \quad (6.33)$$

where boldface is used to indicate vector quantities, since this set of conditions is completely general and the single-nuclear-coordinate consideration is not required. In order to solve this, we can assume that all molecules are equivalent, i.e., they share the same nuclear configuration, their dipoles are aligned, and the coupling strength is the same for all of them. This leads to the single condition $\partial_{\mathbf{R}} V_g(\mathbf{R}) - N\lambda^2\mu_g(\mathbf{R})\partial_{\mathbf{R}}\mu_g(\mathbf{R}) = 0$, for which we get that the change of the equilibrium configuration is given by

$$R_{0,\text{new}} \approx R_0 + N\frac{\lambda^2}{M\omega_\nu^2}\mu_g(R_0)\partial_R\mu_g(R_0) + \mathcal{O}(\lambda^4). \qquad (6.34)$$

This result is reminiscent of the expression in Eq. (6.28), where there is a collective enhancement granted that the remaining molecules have an average global alignment. Note that the single molecule feels the electric field produced by the cavity dipole, which is induced by N aligned dipoles, resulting in a natural enhancement of the electric field of N compared to the single-molecule case. This is similar to the effect of an external electric field, which it is known to strongly impact molecular geometries [68, 69]

Note that the new equilibrium configuration does not depend explicitly on the photon frequency. We analyzed the role of the reaction rates with the frequency in Sect. 6.3, where we see that there are no appreciable changes for high or low photon frequencies. However, we know that within the CBOA, the photon is treated adiabatically, and the method will be more accurate for $\omega_c \to 0$ (e.g., in vibrational strong coupling). Therefore, in the following we study structural changes from a theoretical approach based on the PoPES, used in previous chapters. In this picture, the photonic DoF is discrete, so the electronic–photonic Hamiltonian reads

$$\hat{H}_{e-\text{ph}} = \hat{H}_e + \omega_c\hat{a}^\dagger\hat{a} + \mathbf{E}_{1\text{ph}} \cdot \hat{\boldsymbol{\mu}}(\hat{a}^\dagger + \hat{a}), \qquad (6.35)$$

where $\mathbf{E}_{1ph}$ is the electric field amplitude in the cavity. We now use this Hamiltonian in order to find the ground-state PoPES of the system by doing perturbation theory. The energy shift depends on a sum that runs over states $|i, n\rangle$, where i is the electronic state index and n the photon number. The interaction Hamiltonian $\mathbf{E}_{1ph} \cdot \hat{\boldsymbol{\mu}}(\hat{a}^{\dagger} + \hat{a})$ only allows single-photon transitions, so that in second-order perturbation theory the number of photons is fixed to $n = 1$. This leads to the perturbed ground state

$$\tilde{V}_g(\mathbf{R}) = V_g(\mathbf{R}) - E_{1ph}^2 \sum_i \frac{|\mu_{ig}(\mathbf{R})|^2}{V_i(\mathbf{R}) - V_g(\mathbf{R}) + \omega_c} + \mathcal{O}(E_{1ph}^4). \tag{6.36}$$

We can see that for $i \equiv g$ we get the Debye contribution $-\frac{E_{1ph}^2 \mu_g^2}{\omega_c} = -\frac{\lambda^2}{2}\mu_g^2$ obtained above. We rewrite the ground-state energy as

$$\tilde{V}_g(\mathbf{R}) = V_g(\mathbf{R}) - \frac{\lambda^2}{2}\mu_g^2(\mathbf{R}) - \frac{\lambda^2}{2}\omega_c \sum_{i \neq g} \frac{|\mu_{ig}(\mathbf{R})|^2}{V_i(\mathbf{R}) - V_g(\mathbf{R}) + \omega_c}, \tag{6.37}$$

where we can identify the second term as the usual London dipole–dipole interaction. Note that in the adiabatic limit of the CBOA, where the electronic energies are much larger than the photon frequency, i.e., $V_i(\mathbf{R}) - V_g(\mathbf{R}) \gg \omega_c$, we exactly recover the result of Eq. (6.19). More importantly, since in this picture the surfaces are hybrid electronic–photonic states with parametric nuclear dependence (as opposed to the purely electronic states with parametric photonic–nuclear dependence of the CBOA), the London contribution directly influences the PES instead of the zero-point energy. The new equilibrium configuration therefore includes changes due to the cavity-induced London forces.

It is straightforward to approach the many-molecule scenario by doing perturbation theory on the general Hamiltonian of Eq. (4.4). In this case, it is easy to see that the change in the ground-state PoPES is given by

$$\tilde{V}_g(\mathbf{R}_t) = \sum_i V_g(\mathbf{R}_i) - \frac{1}{2}\left(\sum_i \lambda_i \mu_g(\mathbf{R}_i)\right)^2 - \frac{1}{2}\omega_c \sum_i \sum_{j \neq g} \frac{\lambda_i^2 |\mu_{jg}(\mathbf{R}_i)|^2}{V_j(\mathbf{R}_i) - V_g(\mathbf{R}_i) + \omega_c}. \tag{6.38}$$

Again, this equation is in agreement with the results of Sect. 6.5 and we can therefore extract the same conclusions for the energy changes. In order to analyze the influence on the equilibrium configuration, we assume that all but one molecule are in the equilibrium position $\mathbf{R}_0$. This naturally leads to Eq. (6.33) plus a new term $-\frac{\lambda^2}{2}\omega_c \sum_{j \neq g} \frac{|\mu_{jg}(\mathbf{R}_i)|^2}{V_j(\mathbf{R}_i) - V_g(\mathbf{R}_i) + \omega_c}$. We note that this new term, which represents the London interaction, does not depend on the number of molecules but rather only on the single-molecule coupling. In the limit of a very large number of molecules, while keeping the Rabi splitting Ω_R fixed, the single-molecule coupling goes to zero, and therefore the effect of the cavity London interaction will be negligible.

We thus find that the influence of strong coupling on any specific observable is not immediately obvious, and has to be checked case by case. For some properties, the molecules will behave as if they feel the full collective coupling Ω_R, while for others, they will show only the change induced by the single-molecule coupling λ [5, 70]. These results are also compatible with the experimental observation that the vibrational frequencies in surface-enhanced Raman scattering, which probe the ground-state PES, are not strongly modified under strong coupling [71].

6.7 Conclusions

In this chapter we demonstrated the possibility of modifying ground-state chemical reactions and molecular properties in hybrid cavity-molecule systems without an external input of energy, motivated by experimental results showing this for vibrational strong coupling [1, 4]. By treating a simple model system, the Shin–Metiu model, we were able to show how full thermally driven reaction rates can be significantly modified under strong light-matter interactions. We then determined that this change can be interpreted through classical transition state theory, i.e., by the change in the height of an effective energy barrier (or activation energy) by working within the cavity Born–Oppenheimer approximation. This approximation is particularly accurate for treating vibrational strong coupling, where the cavity frequency is much smaller than the electronic energies. We then use perturbation theory in order to obtain simple analytic expressions relating the effective barrier heights to purely ground-state molecular properties, namely the uncoupled ground-state PES, dipole moment, and polarizability of the molecule. We discuss that within second-order perturbation theory, the energy shifts determining the barrier height on the CBO PES can be directly related to well-known intermolecular forces, i.e., the Debye and London forces, and more generally to Casimir–Polder interactions.

We stress that while perturbation theory allows us to make connections to well-known results, our approach generalizes Casimir–Polder forces beyond the perturbative regime and applies for any coupling strength. Additionally, we have shown explicitly that the emergence of vibrational strong coupling does not affect the validity of the derived expressions for the effective energy barriers. At the same time, the CBOA provides a straightforward way to connect to well-known theories of chemical reactivity. The fact that the energy shifts obtained here become appreciable for realistic nanocavities with strongly sub-wavelength field confinement and thus sufficiently large λ demonstrates that the (generalized) van-der-Waals forces due to the interaction of the molecular dipole with the polarization it induces in the cavity can become strong enough to lead to significant changes in chemical reactivity.

We additionally found that on the single-molecule level, the effects discussed above do not rely on any particular relation between the cavity photon frequency ω_c and the vibrational transitions in the molecule ω_ν, and thus in particular not on the formation of polaritons (hybrid light-matter states). This is consistent with the interpretation of the energy shifts as generalizations of Casimir–Polder interactions beyond the perturbative regime. We also showed that the small modulation of the reaction rate as a function of ω_c that is observed numerically can be understood by simple adiabatic approximations, and again is not related to polariton formation.

We demonstrated the applicability of our approach for a realistic multi-mode cavity, a nanoparticle-on-mirror setup [17], and found that the effective single-molecule coupling strength in this case becomes significant (corresponding to a mode volume of $\approx 2\,\mathrm{nm}^3$) even though the mode volume of the main optically active mode is significantly larger ($\approx 40\,\mathrm{nm}^3$). We furthermore applied our theory to a real molecule, 1,2-dichloroethane, and showed that reaction rates can be both suppressed and enhanced depending on the relative value of the molecular dipole moment at the critical configurations (local minima and saddle points of the PES). A cavity could thus serve as a catalyst or as an inhibitor of a ground-state reaction, and could even alter the global equilibrium configuration of the molecule, all without any kind of external energy input, with all reactions simply driven by thermal fluctuations. This represents a potential way to efficiently optimize the desired yield of a molecular reaction.

For the case of many-molecule strong coupling, where the single-molecule coupling λ is typically so small that the single-molecule effects described above are negligible, we demonstrated that the PES and reaction barriers can be significantly modified by collective effects provided that the permanent dipole moments of the molecules are oriented with respect to the cavity mode field, such that they induce an overall static electric field. However, it should also be noted that similar effects could be achieved by direct dipole–dipole interactions if one manages to align all molecules such as to create a strong field at the position of a single molecule. An interesting open question is whether the cavity-mediated interactions could induce alignment in materials that do not show this in the absence of the cavity, or if direct dipole–dipole interactions would prevent this.

We lastly analyze how the cavity modifies the ground-state nuclear structure. We again find that the Debye and London interactions with the cavity have an impact on the equilibrium nuclear configuration. We furthermore discuss the fundamental limitation in the CBOA for calculating the effect of the London forces on the equilibrium configuration. We thus analyze the same problem from a picture of polaritonic potential energy surfaces, which is consistent with the CBOA approach when the cavity frequency is very small. We find that the influence of the cavity induces collective phenomena for the Debye contribution and single-molecule effects for the London interaction.

Finally, it should be noted that we have throughout assumed that the whole system is in thermal equilibrium, i.e., that the effective temperature is identical both for the molecules and the cavity EM mode. This implies that system-bath interactions do not have to be explicitly modeled, as the system can simply be assumed to be at a given temperature (as explicitly included in the quantum rate calculations and TST). This assumption would break down if the internal vibrational temperature of the molecules is different from the temperature of the thermal radiation bath that the cavity is coupled to. In that case, the effective temperature of the system could potentially become an average of the internal and external bath temperatures. In particular, the effective temperature relevant for a given reaction could depend on whether vibrational motion along that reaction coordinate is hybridized with the cavity mode, such that the external black-body radiation bath would conceivably couple more efficiently to that mode than to others. Such effects have been studied for Casimir–Polder forces, where resonant contributions that exactly cancel at thermal equilibrium can become important in non-equilibrium situations [72, 73], and possibly give rise to additional collective effects [74].

References

1. Thomas A, George J, Shalabney A, Dryzhakov M, Varma SJ, Moran J, Chervy T, Zhong X, Devaux E, Genet C, Hutchison JA, Ebbesen TW (2016) Ground-state chemical reactivity under vibrational coupling to the vacuum electromagnetic field. Angew Chem Int Ed 55:11462
2. Hiura H, Shalabney A, George J (2018) Cavity catalysis -accelerating reactions under vibrational strong coupling-. ChemRxiv 7234721
3. Lather J, Bhatt P, Thomas A, Ebbesen TW, George J (2018) Cavity catalysis by co-operative vibrational strong coupling of reactant and solvent molecules. ChemRxiv 7531544
4. Thomas A, Lethuillier-Karl L, Nagarajan K, Vergauwe RMA, George J, Chervy T, Shalabney A, Devaux E, Genet C, Moran J, Ebbesen TW (2019) Tilting a ground-state reactivity landscape by vibrational strong coupling. Science 363:615
5. Galego J, Garcia-Vidal FJ, Feist J (2015) Cavity-induced modifications of molecular structure in the strong-coupling regime. Phys Rev X 5:41022
6. Martínez-Martínez LA, Ribeiro RF, Campos-González-Angulo J, Yuen-Zhou J (2018) Can ultrastrong coupling change ground-state chemical reactions? ACS Photonics 5:167
7. Flick J, Ruggenthaler M, Appel H, Rubio A (2015) Kohn-Sham approach to quantum electrodynamical density-functional theory: exact time-dependent effective potentials in real space. Proc Natl Acad Sci 112:15285
8. Flick J, Ruggenthaler M, Appel H, Rubio A (2017) Atoms and molecules in cavities, from weak to strong coupling in quantum-electrodynamics (QED) chemistry. Proc Natl Acad Sci 114:3026
9. Flick J, Appel H, Ruggenthaler M, Rubio A (2017) Cavity Born-Oppenheimer approximation for correlated electron-nuclear-photon systems. J Chem Theory Comput 13:1616
10. Angulo JCG, Ribeiro RF, Yuen-Zhou J (2019) Resonant enhancement of thermally-activated chemical reactions via vibrational polaritons. arXiv preprint arXiv:1902.10264
11. Shin S, Metiu H (1995) Nonadiabatic effects on the charge transfer rate constant: a numerical study of a simple model system. J Chem Phys 102:9285
12. Yamamoto T (1960) Quantum statistical mechanical theory of the rate of exchange chemical reactions in the gas phase. J Chem Phys 33:281
13. Miller WH (1974) Quantum mechanical transition state theory and a new semiclassical model for reaction rate constants. J Chem Phys 61:1823
14. Miller WH, Schwartz SD, Tromp JW (1983) Quantum mechanical rate constants for bimolecular reactions. J Chem Phys 79:4889
15. Eyring H (1935) The activated complex in chemical reactions. J Chem Phys 3:107
16. Laidler KJ (1987) Chemical kinetics, 3rd edn. Harper & Row, New York
17. Chikkaraddy R, de Nijs B, Benz F, Barrow SJ, Scherman OA, Rosta E, Demetriadou A, Fox P, Hess O, Baumberg JJ (2016) Single-molecule strong coupling at room temperature in plasmonic nanocavities. Nature 535:127
18. Benz F, Schmidt MK, Dreismann A, Chikkaraddy R, Zhang Y, Demetriadou A, Carnegie C, Ohadi H, De Nijs B, Esteban R, Aizpurua J, Baumberg JJ (2016) Single-molecule optomechanics in "picocavities". Science 354:726

19. Urbieta M, Barbry M, Zhang Y, Koval P, Sánchez-Portal D, Zabala N, Aizpurua J (2018) Atomic-scale lightning rod effect in plasmonic picocavities: a classical view to a quantum effect. ACS Nano 12:585
20. Tokatly IV (2013) Time-dependent density functional theory for many-electron systems interacting with cavity photons. Phys Rev Lett 110:233001
21. Faisal FH (2013) Theory of multiphoton processes. Springer Science & Business Media
22. Del Pino J, Feist J, Garcia-Vidal FJ (2015) Signatures of vibrational strong coupling in Raman scattering. J Phys Chem C 119:29132
23. Shalabney A, George J, Hutchison J, Pupillo G, Genet C, Ebbesen TW (2015) Coherent coupling of molecular resonators with a microcavity mode. Nat Commun 6:5981
24. Shin S, Metiu H (1996) Multiple time scale quantum wavepacket propagation: electron-nuclear dynamics. J Phys Chem 100:7867
25. Abedi A, Agostini F, Suzuki Y, Gross EKU (2013) Dynamical steps that bridge piecewise adiabatic shapes in the exact time-dependent potential energy surface. Phys Rev Lett 110:263001
26. Schneider BI, Feist J, Nagele S, Pazourek R, Hu S, Collins LA, Burgdörfer J (2011) Recent advances in computational methods for the solution of the time-dependent Schrödinger equation for the interaction of short, intense radiation with one and two-electron systems: application to He and H_2^+. In: Bandrauk AD, Ivanov M (eds) Quantum dynamic imaging. CRM series in mathematical physics, vol 149. Springer, New York, NY
27. Feist J (2019) Collection of small useful helper tools for Python. https://github.com/jfeist/jftools
28. González-Tudela A, Huidobro PA, Martín-Moreno L, Tejedor C, García-Vidal FJ (2014) Reversible dynamics of single quantum emitters near metal-dielectric interfaces. Phys Rev B Condens Matter Mater Phys 89
29. Delga A, Feist J, Bravo-Abad J, Garcia-Vidal FJ (2014) Quantum emitters near a metal nanoparticle: strong coupling and quenching. Phys Rev Lett 112
30. Delga A, Feist J, Bravo-Abad J, Garcia-Vidal FJ (2014) Theory of strong coupling between quantum emitters and localized surface plasmons. J Opt 16:114018
31. Li R-Q, Hernángomez-Pérez D, García-Vidal FJ, Fernández-Domínguez AI (2016) Transformation optics approach to plasmon-exciton strong coupling in nanocavities. Phys Rev Lett 117:107401
32. Fried SD, Boxer SG (2017) Electric fields and enzyme catalysis. Annu Rev Biochem 86:387
33. Welborn VV, Pestana LR, Head-Gordon T (2018) Computational optimization of electric fields for better catalysis design. Nat Catal 1:649
34. Bishop DM (1990) Molecular vibrational and rotational motion in static and dynamic electric fields. Rev Mod Phys 62:343
35. Bishop DM (1998) Molecular vibration and nonlinear optics. In: Prigogine I, Rice SA (eds) Advances in chemical physics, vol 1. Wiley-Blackwell
36. Cammi R, Mennucci B, Tomasi J (1998) Solvent effects on linear and nonlinear optical properties of donor-acceptor polyenes: investigation of electronic and vibrational components in terms of structure and charge distribution changes. J Am Chem Soc 120:8834
37. Simpkins BS, Fears KP, Dressick WJ, Spann BT, Dunkelberger AD, Owrutsky JC (2015) Spanning strong to weak normal mode coupling between vibrational and Fabry-Pérot cavity modes through tuning of vibrational absorption strength. ACS Photonics 2:1460
38. Dunkelberger AD, Spann BT, Fears KP, Simpkins BS, Owrutsky JC (2016) Modified relaxation dynamics and coherent energy exchange in coupled vibration-cavity polaritons. Nat Commun 7:13504
39. Casimir HBG, Polder D (1948) The influence of retardation on the London-van Der Waals forces. Phys Rev 73:360
40. Stone A (2013) The theory of intermolecular forces, 2nd edn. OUP Oxford
41. Craig DP, Thirunamachandran T (1998) Molecular quantum electrodynamics: an introduction to radiation-molecule interactions. Courier Corporation
42. Sukenik CI, Boshier MG, Cho D, Sandoghdar V, Hinds EA (1993) Measurement of the Casimir-Polder force. Phys Rev Lett 70:560

43. Buhmann SY (2007) Casimir-Polder forces on atoms in the presence of magnetoelectric bodies. Thesis (PhD), Friedrich-Schiller-Universität Jena
44. Scheel S, Buhmann SY (2008) Macroscopic quantum electrodynamics—concepts and applications. Acta Physica Slovaca 58:675
45. Lombardi A, Schmidt MK, Weller L, Deacon WM, Benz F, de Nijs B, Aizpurua J, Baumberg JJ (2018) Pulsed molecular optomechanics in plasmonic nanocavities: from nonlinear vibrational instabilities to bond-breaking. Phys Rev X 8:011016
46. Becke AD (1993) Density-functional thermochemistry. III. The role of exact exchange. J Chem Phys 98:5648
47. Isborn CM, Luehr N, Ufimtsev IS, Martínez TJ (2011) Excited-state electronic structure with configuration interaction singles and Tamm-Dancoff time-dependent density functional theory on graphical processing units. J Chem Theory Comput 7:1814
48. Ufimtsev IS, Martinez TJ (2009) Quantum chemistry on graphical processing units. 3. Analytical energy gradients, geometry optimization, and first principles molecular dynamics. J Chem Theory Comput 5:2619
49. Titov AV, Ufimtsev IS, Luehr N, Martinez TJ (2013) Generating efficient quantum chemistry codes for novel architectures. J Chem Theory Comput 9:213
50. George J, Shalabney A, Hutchison JA, Genet C, Ebbesen TW (2015) Liquid-phase vibrational strong coupling. J Phys Chem Lett 6:1027
51. De Bernardis D, Jaako T, Rabl P (2018) Cavity quantum electrodynamics in the nonperturbative regime. Phys Rev A 97:043820
52. Axilrod BM, Teller E (1943) Interaction of the van Der Waals type between three atoms. J Chem Phys 11:299
53. Nicosia C, Huskens J (2013) Reactive self-assembled monolayers: from surface functionalization to gradient formation. Mater Horiz 1:32
54. Acuna GP, Möller FM, Holzmeister P, Beater S, Lalkens B, Tinnefeld P (2012) Fluorescence enhancement at docking sites of DNA-directed self-assembled nanoantennas. Science 338:506
55. Chikkaraddy R, Turek VA, Kongsuwan N, Benz F, Carnegie C, van de Goor T, de Nijs B, Demetriadou A, Hess O, Keyser UF, Baumberg JJ (2018) Mapping nanoscale hotspots with single-molecule emitters assembled into plasmonic nanocavities using DNA origami. Nano Lett 18:405
56. Kéna-Cohen S, Davanço M, Forrest SR (2008) Strong exciton-photon coupling in an organic single crystal microcavity. Phys Rev Lett 101:116401
57. Hulliger J, Wüst T, Brahimi K, Martinez Garcia JC (2012) Can mono domain polar molecular crystals exist? Cryst Growth Des 12:5211
58. Hertzog M, Rudquist P, Hutchison JA, George J, Ebbesen TW, Börjesson K (2017) Voltage-controlled switching of strong light–matter interactions using liquid crystals. Chem Eur J 23:18166
59. Waks E, Sridharan D (2010) Cavity QED treatment of interactions between a metal nanoparticle and a dipole emitter. Phys Rev A 82:043845
60. Gonzalez-Ballestero C, Feist J, Moreno E, Garcia-Vidal FJ (2015) Harvesting excitons through plasmonic strong coupling. Phys Rev B Condens Matter Mater Phys
61. Anger P, Bharadwaj P, Novotny L (2006) Enhancement and quenching of single-molecule fluorescence. Phys Rev Lett 96:113002
62. de Vries P, van Coevorden D, Lagendijk A (1998) Point scatterers for classical waves. Rev Mod Phys 70:447
63. Bonin KD, Kresin VV (1997) Electric-dipole polarizabilities of atoms, molecules, and clusters. World Scientific Publishing Co. Pte. Ltd
64. Chipot C, Pohorille A, Castleman AW, Toennies JP, Yamanouchi K, Zinth W (eds) (2007) Free energy calculations. Springer series in chemical physics, vol 86. Springer, Berlin, Heidelberg
65. Mendieta-Moreno JI, Trabada DG, Mendieta J, Lewis JP, Gómez-Puertas P, Ortega J (2016) Quantum mechanics/molecular mechanics free energy maps and nonadiabatic simulations for a photochemical reaction in DNA: cyclobutane thymine dimer. J Phys Chem Lett 7:4391

66. Cortese E, Lagoudakis PG, De Liberato S (2017) Collective optomechanical effects in cavity quantum electrodynamics. Phys Rev Lett 119:043604
67. Keeling J, Kirton PG (2018) Orientational alignment in cavity quantum electrodynamics. Phys Rev A 97:053836
68. Akpati H, Nordlander P, Lou L, Avouris P (1997) The effects of an external electric field on the adatom-surface bond: H and Al adsorbed on Si (111). Surf Sci 372:9
69. Sowlati-Hashjin S, Matta CF (2013) The chemical bond in external electric fields: energies, geometries, and vibrational Stark shifts of diatomic molecules. J Chem Phys 139:144101
70. Ćwik JA, Kirton P, De Liberato S, Keeling J (2016) Excitonic spectral features in strongly coupled organic polaritons. Phys Rev A 93:033840
71. Nagasawa F, Takase M, Murakoshi K (2013) Raman enhancement via polariton states produced by strong coupling between a localized surface plasmon and dye excitons at metal nanogaps. J Phys Chem Lett 5:14
72. Buhmann SY, Scheel S (2008) Thermal Casimir versus Casimir-Polder forces: equilibrium and nonequilibrium forces. Phys Rev Lett 100:253201
73. Ellingsen SA, Buhmann SY, Scheel S (2010) Temperature-independent Casimir-Polder forces despite large thermal photon numbers. Phys Rev Lett 104:223003
74. Sinha K, Venkatesh BP, Meystre P (2018) Collective effects in Casimir-Polder forces. Phys Rev Lett 121:183605

Chapter 7
General Conclusions and Perspective

In this thesis we have focused on studying modifications in the properties and reactivity of organic molecules coupled to cavities hosting confined electromagnetic modes. The aim of these works is to develop a fundamental theory motivated by the various experimental demonstrations of polaritonic chemistry achieved in recent years, both for excited-state molecular processes and thermally-driven ground-state reactions. The work of Chap. 3 is devoted to the first step towards the development of this theory, i.e., combine the usual description of the complexity of organic molecules with theoretical approaches of CQED. We demonstrate the potential of this approach to understand the molecular structure and properties in electronic excited states. The theory is then generalized to an arbitrary number of molecules coupled to a cavity in Chap. 4, where the arising collective phenomena of the system are discussed. Then, in Chap. 5 we present two examples of photochemical reactions that can be manipulated by entering the strong coupling regime. Finally, Chap. 6 is devoted to theoretically study cavity-induced modifications in the ground state, demonstrating the possibility of achieving strong modifications of the molecular energy landscape. In this final chapter we present the overall conclusions of the work developed in this thesis, together with a brief overview of the current status of the related lines of research.

7.1 General Theory of Polaritonic Chemistry

One of the fundamental goals of this thesis is the expansion of our theoretical understanding of polaritonic chemistry. Several chapters focus on establishing a theory that combines CQED and chemistry in order to gain insight of how the cavity can influence the properties of organic molecules. We study both excited and ground states of the hybrid light–matter system with different but related approaches. In

© The Editor(s) (if applicable) and The Author(s), under exclusive license to Springer Nature Switzerland AG 2020
J. Galego Pascual, *Polaritonic Chemistry*, Springer Theses,
https://doi.org/10.1007/978-3-030-48698-3_7

the following, we first review the conclusions related to excited-state phenomena in polaritonic chemistry, and then for the ground-state study, as well as discuss the direction towards a more complete theory of polaritonic chemistry.

In Chap. 3 we explore the use of the Born–Oppenheimer approximation in a hybrid light–matter system. We find that this is still a good approach in electronic strong coupling and can be used to generalize the concept of PES to polaritonic PES (PoPES). This picture is a useful platform to study the chemical properties in organic polaritons and to understand the nature of their various nonradiative processes in terms of nuclear relaxation on the PoPES and through nonadiabatic transitions between different polaritonic states. The description is further formalized in Chap. 4, where an extension to treat an arbitrary number of molecules and excitations is presented. This allows to study the different collective phenomena that arise in strong coupling with an ensemble of molecules, such as the collective conical intersections and the collective protection effect. The latter phenomenon is crucial in the modification of the excited-state structure of the system and thus the primary responsible of influencing the different photochemical reactions studied in Chap. 5. This chapter is focused on the study of some possible modifications of photochemistry, such as suppression of excited-state processes, and triggering of multiple photochemical reactions on many molecules after photoabsorption of a single external photon.

Throughout this thesis we establish the potential of the PoPES picture for studying various excited-state processes, providing specific examples of photochemical reactions with simplified molecular models. This theory is very flexible and its principles allow to increase the complexity of both the electromagnetic and molecular components. For instance, this theory can be interfaced with state-of-the-art quantum chemistry codes in order to make accurate predictions on particular reactions. Various works show this for different molecules and computational methods, such as the multi-configuration time-dependent Hartree method for propagation of multi-dimensional wavepackets [1] and an on-the-fly surface hopping semiclassical technique for calculating the dynamics on the PoPES [2, 3]. In the work of Luk et al. [2], rhodamine molecules were characterized in detail using a quantum mechanics/molecular mechanics (QM/MM) approach, where the most relevant part of the big molecule for characterization of the photophysics was treated quantum mechanically, while the remaining part, as well as the molecular solvent, was described through classical molecular dynamics. This description allows to fully account for nonradiative losses of the excitation, while spontaneous decay can be accounted for by including stochastic jumps to the ground state.

The inclusion of various coherence and excitation loss mechanisms is an important goal in a general theory of polaritonic chemistry due to the important role that disorder, decay, and decoherence play in organic polaritonics. While these can arise naturally from a detailed-enough microscopic theory (see for example the QM/MM approach mentioned above, or using a detailed quantum electrodynamical density functional theory approach [4, 5]), this rapidly becomes computationally infeasible as the system grows in size, and more sophisticated open-quantum system theories are required. Previous attempts to combine the picture of standard PES with losses have been made. For example, PES can be generalized to complex PES [6], where the imaginary part represents the decay in time of the state. Thus, both real and imaginary PES surfaces can intersect, leading to a variety of novel effects. Coupled system–bath approaches can also be used to simulate the effect of weakly coupled vibrational degrees of freedom such as the environment of a molecule in a condensed phase. In this context, quantum chemistry methods have been combined with Redfield theory to compute the nonadiabatic photochemical dynamics of the pyrrole-pyridine hydrogen-bonded complex [7]. Analogous approaches can readily be implemented in polaritonic chemistry in order to fully understand the role of losses in these processes. The final objective of a general theory of polaritonic chemistry is to fully integrate the methods and understanding of both chemistry and QED to develop a unified insight on how to modify chemical properties in organic molecules.

Finally, we devote Chap. 6 to the theoretical study of cavity-modified ground-state molecular structure and reactivity. In this chapter we thoroughly discuss chemical changes in terms of the cavity Born–Oppenheimer approximation. This approach is related to the picture of PoPES, and therefore offers similar advantages. The microscopic theory that we develop connects the predicted changes in chemical structure and reactivity with off-resonance Casimir–Polder interactions, and does not require the formation of polaritons. We note that current experiments of cavity-modified ground-state chemistry observe that the cavity effects are resonance-dependent [8–11]. At the time of writing this thesis, only another theoretical work has approached this problem [12]. This work finds a resonant condition for a specific theoretical model where only increments in reactivity can be obtained, however the experiments observed both catalysis and suppression of chemical reactions. The mechanism responsible for the experimentally observed resonance-dependent changes in reactivity thus remains unknown, and further theoretical work is needed to uncover it. More complete descriptions are required to understand these phenomena, by, e.g., explicitly including the role of the solvent (which is known to play a role [10]), or using a thermodynamical description that does not necessarily assume thermal equilibrium between cavity and molecules.

Nevertheless, our theory provides a platform to manipulate chemistry in plasmonic nanocavities. In Chap. 6 we provide examples for realistic molecules, using quantum chemistry packages to compute molecular properties, and for realistic cavities, exploiting electrostatics to fully take into account all the EM modes coupled to the molecule. Our latest work [13], not featured in this thesis, focuses on studying various chemical systems, demonstrating cavity-induced catalysis and control over the spin states and transition temperature in spin-crossover complexes. We believe that readily available cavities, such as the nanoparticle-on-mirror [14], could be exploited to achieve the few-molecule chemical control predicted by our theory.

7.2 Applications of Cavity-Modified Chemistry

Due to the infancy of the field of polaritonic chemistry, there is still a need to apply this theory to more realistic and bigger systems, which, as discussed above, will require combining it with additional techniques to achieve a more complete description. Accordingly, most of this thesis has been focused on establishing the appropriate theoretical framework. In Chap. 5 we use this theory to predict and describe different effects that may be exploited in experiments, and, in the long term, in possible technological applications. Particularly, we describe the effect of collective suppression in strong coupling, which leads to important rate reductions in excited-state processes. Some recent studies have demonstrated experimentally that this mechanism can significantly influence the intersystem crossing rates between singlet and triplet states [15, 16]. In the triplet state, interactions with external triplet oxygen can lead to bleaching of the molecule. Therefore, this suppression can help with stabilization of highly reactive molecules, reducing photodegradation both in plasmonic systems [15] and in planar cavities [16].

Additionally, in Chap. 5 we presented a proof-of-principle study showing the possibility of triggering several photochemical reactions among many molecules with a single external photon. This is possible for the molecular model treated, which displays an energy landscape typical in organic molecules used for energy storage in solar cells [17]. This energy is stored in a metastable nuclear configuration with a very long lifetime, thus making this system good for storing energy, but not for retrieving it. In our study we demonstrate that, in strong coupling, this same molecular system can be good for retrieving energy and not for storing it, as photoabsorption of a single photon can trigger multiple energy-releasing back-reactions. By reversibly bringing the system in and out of resonance, by, e.g., moving a mirror that changes the photon frequency, it could be feasible to control whether the system stores energy or releases it.

The nonadiabatic dynamics in organic polaritons that we explore in Chaps. 3 and 4 also offer a wide range of possibilities. The nonadiabatic transitions and conical intersections that are induced in strong coupling are closely related to the so-called light-induced conical intersections in strong laser fields [18]. The use of light to steer molecular dynamics is the main goal of the field of coherent control [19], in which specifically tailored EM fields play the role of "photonic reagent" or "photonic catalyst". This approach accomplish strong interactions with large number of photons n, exploiting the enhancement of the laser electric field amplitude as $\sim \sqrt{n}$. In strong coupling, the basic properties of this phenomenon are instead achieved using the large strength of a single photon. Therefore, polaritonic chemistry can considerably learn from the field of coherent control, which has enabled manipulation of the dynamics of several molecular reactions [20, 21] and control over dissociation in molecules [18, 22]. Polaritonic chemistry can potentially offer similar chemical control, while avoiding some of the drawbacks of strong laser physics, such as undesired multiple photoabsorptions and ionizations that often limit the possible utility of coherent control.

In the next few years the potential of polaritonic chemistry will surely be further explored. The possibility of manipulating the chemical and material properties of organic systems offer plenty of opportunities. For example, it could offer the catalysis of reactions for which no conventional catalytic method is known, or to do so cheaper and more efficiently than in standard catalysis. This capability could also lead to deeper understanding of many important processes in nature, such as the photophysics in photosynthesis and human vision. Nature often controls these systems by manipulating the energy landscape of the molecules with a surrounding protein [23]. Strong coupling could offer a novel alternative to manipulate these crucial processes. Furthermore, the robustness of organic polaritons opens possibilities towards room-temperature quantum technologies. Polaritonic chemistry would offer an additional tool to control and tune the properties of organic systems in order to optimize its use in technological applications.

7.3 Ending Remarks

The field of polaritonic chemistry has demonstrated to be an interesting topic that has attracted increasingly more interest in the past recent years. It offers a novel approach for modifying the properties of organic systems by using cavities to alter the electromagnetic vacuum that dresses and defines the molecules. This is still a very young field where more experimental studies are required in order to verify and challenge current theoretical predictions. At the same time, theory must provide the necessary tools to understand these experiments, by allowing the treatment of more complex descriptions of the system. In this context, this thesis has been focused on addressing this problem, by providing a satisfactory theoretical method to describe organic polaritons. We build on the foundations of chemistry and QED in order to provide an initial step towards a more complete theory of polaritonic chemistry. In the long term, a goal of the field will certainly be to investigate the use of cavity-modified chemistry for practical applications.

The interdisciplinary aspect of polaritonic chemistry is one of its greatest strengths. Nanophotonics and quantum chemistry are combined in this emergent field with the promise of great fundamental and technological development. As different scientific communities notice this innovative concept, the influx of new ideas will ensure a boost of our understanding of cavity-modified chemistry. A new generation of quantum technologies and creative ways to manipulate complex chemical systems could be possible by exploiting the characteristics of light in the dark.

References

1. Vendrell O (2018) Coherent dynamics in cavity femtochemistry: application of the multi-configuration time-dependent Hartree method. Chem Phys 509:55
2. Luk HL, Feist J, Toppari JJ, Groenhof G (2017) Multiscale molecular dynamics simulations of polaritonic chemistry. J Chem Theory Comput 13:4324
3. Fregoni J, Granucci G, Coccia E, Persico M, Corni S (2018) Manipulating azobenzene photoisomerization through strong light-molecule coupling. Nat Commun 9:4688
4. Schäfer C, Ruggenthaler M, Rubio A (2018) Ab initio nonrelativistic quantum electrodynamics: bridging quantum chemistry and quantum optics from weak to strong coupling. Phys Rev A 98:043801
5. Flick J, Welakuh DM, Ruggenthaler M, Appel H, Rubio A (2018) Light-matter response functions in quantum-electrodynamical density-functional theory: modifications of spectra and of the Maxwell equations. arXiv preprint arXiv:1803.02519
6. Feuerbacher S, Sommerfeld T, Cederbaum LS (2004) Intersections of potential energy surfaces of short-lived states: the complex analogue of conical intersections. J Chem Phys 120:3201
7. Lan Z, Frutos LM, Sobolewski AL, Domcke W (2008) Photochemistry of hydrogen-bonded aromatic pairs: quantum dynamical calculations for the pyrrole-pyridine complex. Proc Natl Acad Sci 105:12707
8. Thomas A, George J, Shalabney A, Dryzhakov M, Varma SJ, Moran J, Chervy T, Zhong X, Devaux E, Genet C, Hutchison JA, Ebbesen TW (2016) Ground-state chemical reactivity under vibrational coupling to the vacuum electromagnetic field. Angew Chem Int Ed 55:11462
9. Hiura H, Shalabney A, George J (2018) Cavity catalysis -accelerating reactions under vibrational strong coupling-. ChemRxiv 7234721
10. Lather J, Bhatt P, Thomas A, Ebbesen TW, George J (2018) Cavity catalysis by co-operative vibrational strong coupling of reactant and solvent molecules. ChemRxiv 7531544
11. Thomas A, Lethuillier-Karl L, Nagarajan K, Vergauwe RMA, George J, Chervy T, Shalabney A, Devaux E, Genet C, Moran J, Ebbesen TW (2019) Tilting a ground-state reactivity landscape by vibrational strong coupling. Science 363:615
12. Angulo JCG, Ribeiro RF, Yuen-Zhou J (2019) Resonant enhancement of thermally-activated chemical reactions via vibrational polaritons. arXiv preprint arXiv:1902.10264
13. Climent C, Galego J, Garcia-Vidal FJ, Feist J (2019) Plasmonic nanocavities enable self-induced electrostatic catalysis. Angew Chem Int Ed
14. Chikkaraddy R, de Nijs B, Benz F, Barrow SJ, Scherman OA, Rosta E, Demetriadou A, Fox P, Hess O, Baumberg JJ (2016) Single-molecule strong coupling at room temperature in plasmonic nanocavities. Nature 535:127
15. Munkhbat B, Wersäll M, Baranov DG, Antosiewicz TJ, Shegai T. Suppression of photo-oxidation of organic chromophores by strong coupling to plasmonic nanoantennas
16. Peters VN, Faruk MO, Asane J, Alexander R, D'angelo AP, Prayakarao S, Rout S, Noginov M (2019) Effect of strong coupling on photodegradation of the semiconducting polymer P3HT. Optica 6:318
17. Kucharski TJ, Tian Y, Akbulatov S, Boulatov R (2011) Chemical solutions for the closed-cycle storage of solar energy. Energy Environ Sci 4:4449
18. Corrales M, González-Vázquez J, Balerdi G, Solá I, De Nalda R, Bañares L (2014) Control of ultrafast molecular photodissociation by laser-field-induced potentials. Nat Chem 6:785
19. Brif C, Chakrabarti R, Rabitz H (2010) Control of quantum phenomena: past, present and future. New J Phys 12:075008
20. Kim J, Tao H, White JL, Petrovic VS, Martinez TJ, Bucksbaum PH (2011) Control of 1, 3-cyclohexadiene photoisomerization using light-induced conical intersections. J Phys Chem A 116:2758
21. Madsen CB, Madsen LB, Viftrup SS, Johansson MP, Poulsen TB, Holmegaard L, Kumarappan V, Jørgensen KA, Stapelfeldt H (2009) Manipulating the torsion of molecules by strong laser pulses. Phys Rev Lett 102:073007

22. Sussman BJ, Townsend D, Ivanov MY, Stolow A (2006) Dynamic Stark control of photochemical processes. Science 314:278
23. Domcke W, Yarkony DR, Köppel H (2011) Conical intersections: theory, computation and experiment, vol 17. World Scientific

Printed in the United States
by Baker & Taylor Publisher Services